# Smart Prisons

Peiliang Sun

# Smart Prisons

華中科技大學出版社
Huazhong University of Science and Technology Press

Springer

Peiliang Sun
Hangzhou, Zhejiang, China

ISBN 978-981-16-9656-5 ISBN 978-981-16-9657-2 (eBook)
https://doi.org/10.1007/978-981-16-9657-2

Jointly published with Huazhong University of Science and Technology Press
The print edition is not for sale in China (Mainland). Customers from China (Mainland) please order the print book from: Huazhong University of Science and Technology Press.



This Springer imprint is published by the registered company Springer Nature Singapore Pte Ltd.
The registered company address is: 152 Beach Road, #21-01/04 Gateway East, Singapore 189721, Singapore

# Preface

The rapid development of the new generation of information technologies, such as the IOT, cloud computing and big data, is triggering a new round of industrial transformation around the world. Economic and social development is driven to run at an unprecedented speed, and people's lifestyle has changed greatly. By utilizing perceptual technology and intelligent processing technology in IOTs, the physical world can be recognized, analyzed, processed and subjected to knowledge mining. It realizes the interaction and seamless link between human beings and things and between two or more things. The relationship between humans and nature is comprehensively changed. Big data originates from the large-scale application of the IOT, characterized by distributed data mining of massive data. It is supported by the distributed processing of cloud computing, distributed databases, cloud storage and virtualization technology. Data collected through IOT can be analyzed and processed in cloud computing and big data in a real-time manner. Meaningful and useful rules and results can be obtained by completion of complicated computing and complex controls in massive amounts of data, which forms the basis of intelligent and considerate services. Human activities will become simpler and more delicate with the incoming era of intelligence.

The prison, as the national penalty execution institution, is granted with the responsibilities of maintaining social peace and stability and preventing and reducing crimes. To ensure the fulfillment of its duties, the support of modern information technology is indeed an integral part. Just as the concept of smart cities with IOT technology as the core provides a new model for future urban development, smart prison is also the direction of future prison development. The construction of smart prisons can promote the revolution of prison infrastructure and management. The reform of prison security, correction, law enforcement and administration are enforced with the implementation of strengthening public security through science and technology. It is not only the innovation of prison management but also the inevitable and fundamental requirement of modern civilized prisons.

With the incoming intelligent era, new challenges of prison management and law enforcement have attracted attention. To promote the development of prison management and law enforcement, major problems urgently need to be solved, including state-of-the-art design, construction, standards and evaluation of smart prisons, the protection of privacy in prison clouds and big data processing. Great efforts have been made to search for feasible and suitable solutions to the problems mentioned above.

The author has been committed to applying the new generation of information technology to the research and practice of prison management for many years. As the first pioneering work of "Smart Prison", it provides a reference for the construction of intelligent and civilized prisons. Technologies mentioned in the book, including IOT, cloud, big data and so on, outline the comprehensive and clear development of smart prisons, which will play a positive role in the research and construction of further smart prisons.

Hangzhou, China Peiliang Sun

# Description

Based on prison-related themes such as the Internet of Things (IOT), cloud and big data, this book systematically analyzes and studies smart prisons in terms of the overall architecture, construction paths and intelligent applications. This is the first time that the state-of-the-art design and the evaluation index of smart prisons have been put into specific detail. The aim of this book is to provide readers with a clear and full understanding of the current prison informatization and intelligent development trend. A feasible solution for the informatization and modernization of prison governance can be further explored and researched.

This book is intended to be taken as professional working guidance for staff engaged in prison, drug rehabilitation centers, detention houses and community correction institutions. It is also suitable for use as a teaching/training/reference book for college students majoring in criminal execution, prison management, community correction, judicial information technology, judicial information security etc.

# Contents

## Part II Intelligent Security Protection



## Part III Smart Application

# Part I
# State of the Art Design

# Chapter 1
# Smart Prison State of the Art Design

There is a new concept in how we view what "smart" is. IBM's "smarter planet" is designed to change the way humans live their daily lives by integrating the physical with the digital through embedded sensors to provide connectivity between humans and the various smart systems that we use today. The aim is, with the use of supercomputers and cloud computing providing integration to the internet of things, to achieve complete analysis of all we do via the utilization of large data, thus promoting productivity and improving our way of life.

At a time when change at all levels has promoted a major shift in social governance by means of statistically driven data decision making, from passive reaction to active foresight, from extensive management to fine, individualized transformation, the implementation of big data strategy-led organizational management changes and business model innovation, has become the new normal development of this era.

Led by this new concept of "smart", opportunities abound for the revamping of prison operations and correction models by utilizing innovations, such as the Internet of Things and big data, which represent the next generation of information technologies. Through this, reforms to prison systems can be achieved by providing a platform that meets the expectation of social and economic development and the creation of modern, more state of the art prisons. In turn, this would ensure proper supervision at all levels, improve the quality of education and rehabilitative services, promote fair enforcement of the law and improve the standards of correctional officers.

To fully realize overall functionality, the value of prison supervision, and the rehabilitation of criminals, a smart prison needs the foresight of prevention and control and refined management and fairness at all levels of law enforcement. Decision making must be precise, emphasizing the overall perception of prisons and their associated parts. This is especially so under the purview of its inmate-oriented concept of intelligence-driven correction by the establishment of a corrections knowledge base.

A smart prison relies on an advanced form of prison informatization. The first step is the new generation of information technology, which is represented by the Internet



P. Sun, *Smart Prisons*,
https://doi.org/10.1007/978-981-16-9657-2_1

of Things, cloud computing and big data. The second is the innovative ecology of the prison work management model, and the third is the goal of a modern state of the art prison.

The core issues considered and discussed in this book are how to successfully design a state of the art smart prison by applying high-tech solutions such as the Internet of Things, cloud computing, big data and artificial intelligence. The goal is to enhance prison security technology prevention capabilities, improve inmate education and rehabilitation, and improve corrections management and the overall standard of corrections personnel performance.

The first chapter will focus on prison informatization by utilizing state of the art architecture design concepts to refine supervisory methods and the overall transformation of prison operations. This will provide clarity in the key aspects of prison work and integrated prison administration management which includes penalty enforcement, rehabilitation, inmate work details, inmate recreation and health. This state of the art design would also incorporate the management of prison staff and other core business requirements.

## 1.1 Architectural Methodology

State of the art design in this case refers to the use of system theory methods from a global perspective. This covers a task or project from all aspects, levels and elements of the planning process in its entirety, allowing the effective allocation of resources and efficient realization of targets. As prisons are large complex systems involving multiple elements at various levels, including engineering, business, applications, data and technology among numerous others, it is necessary to properly coordinate these elements to ensure the rational planning and effective operation of the prison information system.

This state of the art design consists of three main components: information resource planning, technical architecture planning and environmental security planning. Information resource planning is proposed based on mainstream business analysis. This is achieved by dividing functional domains according to business, conducting business analysis on functional domains one by one, generating functional models and data models, analyzing management models and user behaviors, and generating user models. The purpose of technical architecture planning is to formulate software architecture schemes and application system architecture schemes based on the status quo and existing problems of informatization. Environmental security planning plans and arranges the network, server architecture, storage and backup, computer room design, and information security operation and maintenance to ensure the normal development of informatization construction and the safe and normal operation of the system (Fig. 1.1).

Prison informatization and smart prisons are not only simple applications of information technology but also include the transformation of prison management

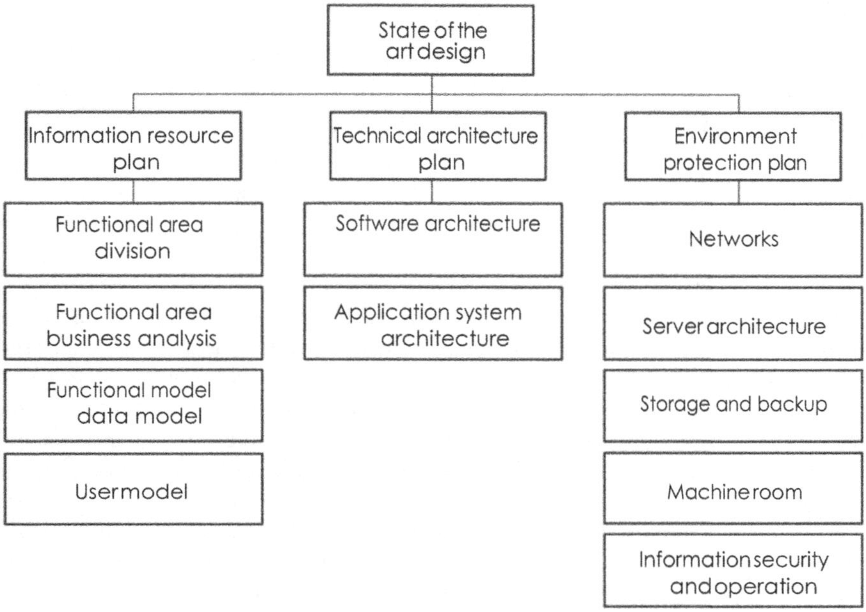


**Fig. 1.1** Contents of the state of the art design

models that require systematic methodological guidance. EA (enterprise architecture) methodology was produced under this background and has been developed rapidly. In 1987, Zachman proposed the first architecture framework at the system architecture level, the Zachman framework. Later, this was developed into a classic framework of EA, and Zachman is also known as the Father of EA. The current international mainstream EA frameworks and methodology include the Top Open Group Achitecture Framework (TOGAF), US Federal System Architecture (FEA), and the US Department of Defense Framework (DoDAF), etc.

The methodology for this state of the art design combines information-based theory and practice. Therefore, the outcome is a blueprint for prison informatization and smart prison construction under the guidance of planning, design business architecture, data architecture, application architecture and technical architecture. This lays out the roadmap for informatization and intelligent construction within the next 3–5 years and sets the path to achieve the target outcomes (Table 1.1).

The Open Group, an international standard authority organization, began to formulate system architecture standards in response to customer requirements in 1993 and released the Open Group Architecture Framework (TOGAF) in 1995. The basis of TOGAF is the Technical Architecture Framework for Information Management (TAFIM) of the US Department of Defense. It is based on an iterative process model

**Table 1.1** The main EA framework comparison

| Framework | The first edition | The latest version | Integrity | Practicability |
|---|---|---|---|---|
| Zachman | 1987 | 2003 | Mid | Low |
| EAP | 1992 | No | Mid | Mid |
| FEA | 1999 | Updated all the time | High | High |
| DoDAF | 2003 | 2009 | High | High |
| TOGAF | 1995 | 2009 | High | High |

and is a tool that assists in the development, acceptance, operation, use and maintenance of the architecture. TOGAF is currently one of the most mainstream organizational structure methods in the world. It has been used by 80% of Forbes 50 companies and supports an open and standard SOA reference architecture. It has been proven to be a flexible and efficient way to build enterprise IT architecture. Effective methodology is currently being promoted mainly by companies such as SAP, IBM, HP, and SUN.

This chapter mainly adopts the TOGAF state of the art design method and is appropriately tailored according to the actual situation of prisons. TOGAF covers four basic types of architecture, as shown in the figure below (Fig. 1.2).

◎ Business architecture: business strategy, governance, organization, and key business process information and their interactions;

◎ Data architecture: the structure of the organization's various logical and physical data assets and data management resources;

◎ Application architecture: description of the individual application systems deployed, the interactions between systems, and their relationship to the organization's core business processes;

◎ Technology architecture: hardware and software capabilities that are necessary to support the deployment of business, data and application services, including IT infrastructure, middleware, networking, communications, deployment processing and standards.

The overall structure of the smart prison will be designed from these four categories accordingly.

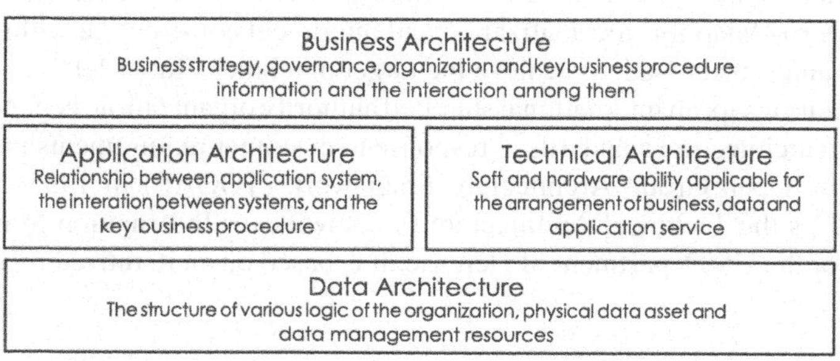


**Fig. 1.2** EA category 1

## 1.2 Status and Gap

At present, China's prison information program has made a series of important achievements, but there are also some problems that cannot be ignored, such as inadequate top-level planning, repeated construction investment, and information silos.

◎ Business level: Information construction still has difficulty meeting the needs of prison business management innovation and rapid development.

◎ Data level: information resources are dispersed, data resource consistency is difficult to ensure and information sharing is difficult. Information silos are common across departments and business synergy is difficult.

◎ Application level: application system repeated construction, application depth and breadth is insufficient. The application potential is not fully realized, the supporting effect of prison management innovation is weak and the support for scientific decision-making is limited.

◎ Infrastructure level: insufficient information performance evaluation, repeated investment in hardware and software construction, wastage of resources and the generally overlong life of infrastructure equipment performance makes it difficult to support the deployment of complex applications.

Therefore, by making full use of emerging technologies such as the Internet of Things, cloud computing, big data, and artificial intelligence, through structural reorganization, process reengineering and data integration, a unified informatization and intelligent prison construction model can be established. This will avoid investment waste and information silos, serve to solve the contradiction between the content of informatization construction and the actual business needs of prisons as well as between the application of informatization and the actual management needs of prisons to create a new ecology in which information technology empowers prison governance.

◎ To construct a decision-making system of prison risk assessment supported by data and realize the scientific management of prisons.

◎ Let the "data run more" to help prison governance quality and efficiency and to achieve fine governance.

◎ Breaking down data barriers, eliminating information silos, promoting cross-sectoral, cross-level linkage and achieving collaborative governance.

## 1.3 Smart Prison Business Structure

### *1.3.1 Component Business Model*

The component business model (CBM) is a business modeling technique proposed by IBM that divides the business into nonoverlapping business components, helping us look at the business at many different levels and enabling component analysis across business lines.

According to the CBM business capability model and combined with prison business characteristics and business reality, this defines the two dimensions of "responsibility level" and "business capability" and the breakdown of business components.

(1) Level of responsibility—classification of prison functions

Functional classification is used to divide the scope and purpose of business decisions. The "state of the art design" of prisons serves operational work at all levels of responsibility in prisons and can therefore be classified as follows (Fig. 1.3):

◎ Business decision-making level

From the perspective of the prison service and prison leaders, the overall strategy of the prison, the system construction and the planning decision-making work of the key business areas of concern at this level are reviewed.

◎ Business management level

From the perspective of the Bureau of Prisons and the middle-level managers of prisons, we sift through the business management categories of concern at this level and are responsible for handling important and critical matters.

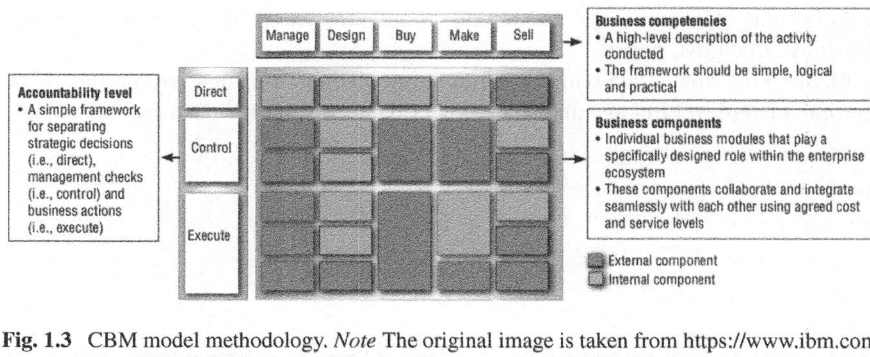


**Fig. 1.3** CBM model methodology. *Note* The original image is taken from https://www.ibm.com/downloads/cas/8RB5RWVE, Pohle, Korsten P, Ramamurthy S (2005) Component business models: making specialization real. Whitepaper issued by IBM institute for business value. IBM Global Services. Somers, New York

◎ The business operations hierarchy

From the perspective of all kinds of business-specific operators, review the specific work content of all aspects of prison operations.

The functional classifications at the three levels of operational decision-making, operational management and operations are reflected in the three organizational levels of the prison service, prison's and subdistricts, rather than simply by correspondence.

The relationship of supervision and guidance between supervisors and staff of the prison system are mainly reflected on two levels:

First, at the macro level, the guidance and supervision of the superiors on the general problems that arise at the lower levels and on the practical difficulties that the lower levels need to address, that is, on all duty-related activities of all prison personnel.

Second, at the micro level, superiors provide guidance on specific problems or certain types of matters in some prisons, and the content of such guidance and supervision is often determinative and directional.

As a result, the three functional classification effectiveness areas of operational decision-making, business management and operations between prison organizational levels and the relationship between superiors and staff are shown in the following figure (Fig. 1.4):

(2) Operational capacity—prison business subject

The prison business subjects are used to define the scope of the prison business categories covered by the "state of the art design", reflecting this design support for the prison business.

According to the actual business situation of the prison, business subjects can be divided into the following categories: supervision security, penalty correction, fair law enforcement, comprehensive logistic support, team building, and business coordination.

◎ Supervision security category

Ensuring the security and stability of the prison is the basis for its successful operation and is the top priority of prison work. The supervision of security includes the business activities of prison case investigation, duty patrol and monitoring, equipment review for misuse prevention, emergency response command and dispatch as well as management decision-making and business guidance for all levels of supervision and security.

◎ Penalty correction category

Penalty correction is the most important objective of the prison and it includes prison administration, inmate education, inmate job opportunities and other corrective penalties carried out in accordance with the law.

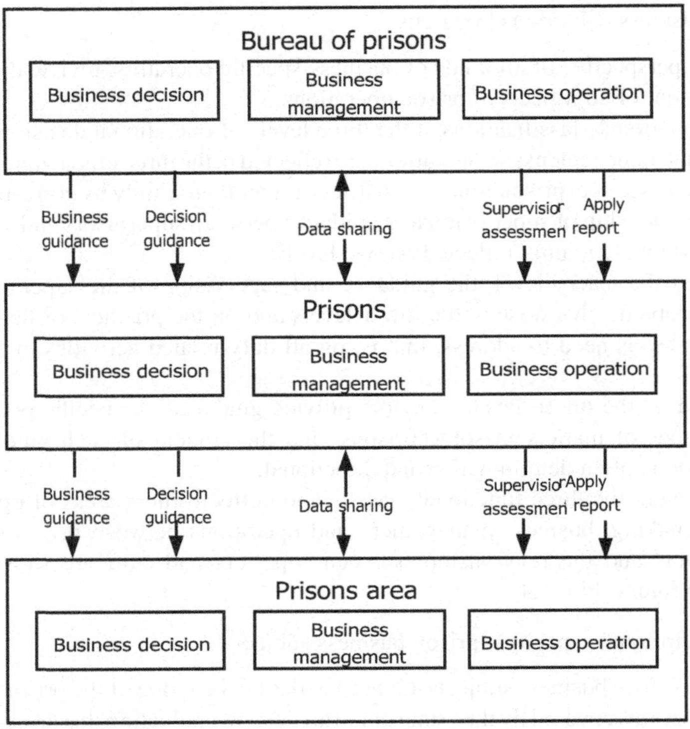


**Fig. 1.4** Prisons three organizational level—functional classification diagram

◎ Fair law enforcement category

Ensuring justice, civility and integrity of law enforcement is an important aspect of ensuring the correct execution of penalties in prisons and maintaining the seriousness of national laws, including the disclosure of prison services complaints, civilian services, legal aid for inmates, publicity of various policies and regulations, letters and visits from relatives, management decision-making and business guidance at all levels of fair law enforcement operations.

◎ Comprehensive logistic support category

Comprehensive logistic support assists the prison penalty correction program in carrying out necessary support, which includes health services, material and equipment distribution, financial management, information services, logistics management, administrative file management, infrastructure construction and other business activities, as well as comprehensive logistic support plans at all levels of management decision-making and business guidance.

◎ Team building category

Team building refers to the relevant content of personnel team building in prisons, including political and police personnel programs, performance appraisal programs, education and training programs, supervision and audit programs and other activities. This also includes management decision-making and business guidance for all levels of team building programs.

◎ Business synergy category

Business synergies are operational linkages and information-sharing activities within prisons. These are between prisons and prison administrations, prisons and external units, as well as management decision-making and operational guidance synergies at all levels of operation. External units include courts, procuratorates, public security, judicial administration, etc.

(3) Business components—prison business functions

Business components are used to deliver or provide operational functions at the appropriate prison management levels to achieve the appropriate hierarchical objectives. Ultimately, all operational components form operational capabilities to jointly achieve the overall strategic objectives of the prison.

A business component is a series of business actions with in-house connections, with the ability to operate independently and each business component has:

◎ The ability to provide unique operational functions for prisons;
◎ The ability to work with other internal business components and external business services;
◎ The ability to operate on the basis of coordinated workflows;
◎ The ability for operating results to be managed and assessed.

### *1.3.2 Business Architecture*

In the state of the art design of smart prisons, business architecture focuses on prison business and its processes. Regulatory security services are classified according to the functional classification of business decision-making level, business management level, and business operation level. The corresponding business components (prison business functions) are:

Business decision-making level: prison security supervision system, prison security confidentiality system, information security equipment use and maintenance system, emergency plan management system, etc.

Business management level: emergency response management, security on duty management, key monitoring criminal management, criminal activity content management, criminal activity venue management, etc.

Business operation level: prison case investigation, duty patrol and monitoring, equipment misuse investigation and prevention, emergency response, command and dispatch, etc.

The regulatory security business architecture is shown in Fig. 1.5.

Penalty correction is classified according to the functional classification of business decision-making level, business management level and business operation level. The corresponding business components (prison business functions) are:

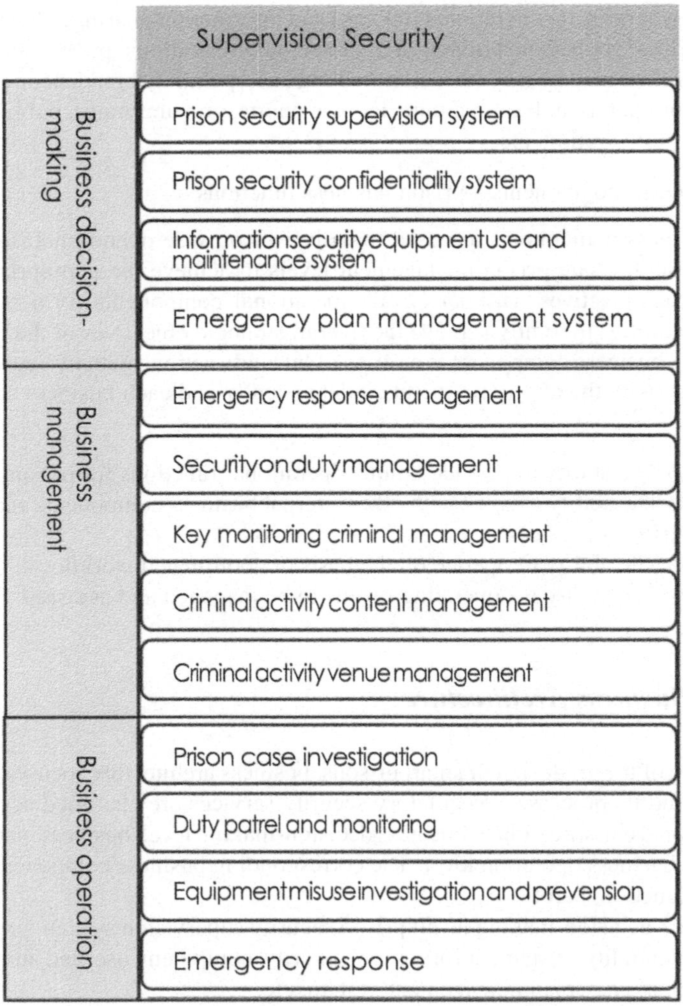


**Fig. 1.5** Overseeing the security business architecture

Business decision-making level: Penalty correction quality management system, penalty correction efficiency control system, penal correction work system standard, penal correction law research, penalty correction business statistics, prison situation analysis, etc.

Business management level: Penalty correction quality management, penalty correction efficiency management, penalty correction process management, penalty correction level management, penalty correction performance management, etc.

Business operation level: Prison administration, penalty enforcement (entering and leaving prison, commutation and parole, temporary furlough outside of the prison, etc.), education correction, labor production, etc.

The panel correction business structure is shown in Fig. 1.6.

**Fig. 1.6** Penalty corrections business structure

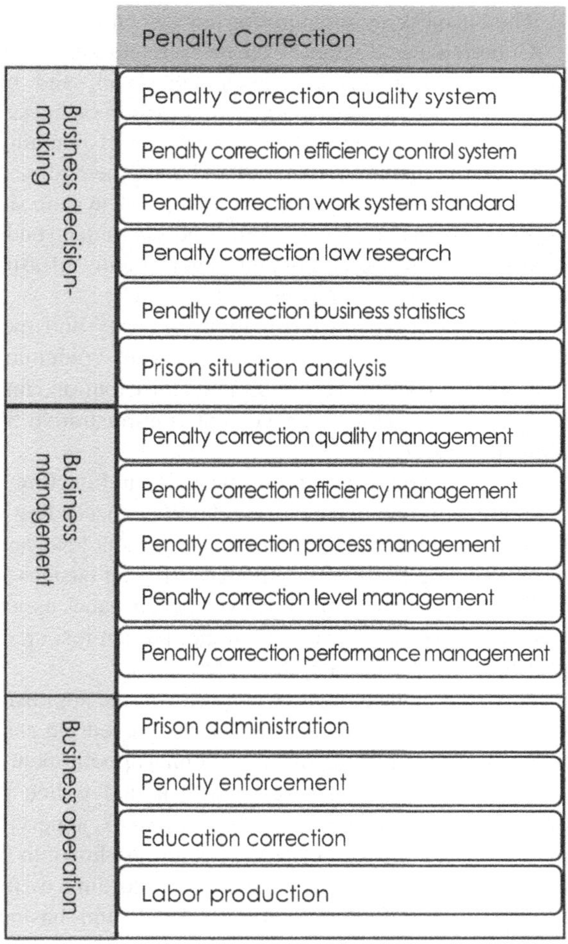

Fair law enforcement services are classified according to the functions of business decision-making level, business management level and business operation level. The corresponding business components (prison business functions) are:

Business decision-making level: prison affairs disclosure system, prisoners' rights protection system, rule of law publicity system, convenience measures system, etc.

Business management level: service channel management (self-service terminal, telephone, network), service process management, service equipment management, service resource management, etc.

Business operation level: public prison affairs (public prison affairs for prisoners, public prison affairs for prisoners' relatives, etc.), complaints and reports, convenience services (self-service terminals, telephone, Internet, remote video interviews, etc.), prison rules and regulations publicity, policies, regulations and rule of law publicity, legal aid to inmates, letters and visits from relatives of inmates, etc.

The business architecture for fair law enforcement is shown in Fig. 1.7.

Comprehensive support business is classified according to business decision-making level, business management level, and business operation level. The corresponding business components (prison business functions) are:

Business decision-making level: overall planning for fund use, infrastructure construction planning, material and equipment allocation planning, informatization construction planning, administrative work system standardization, etc.

Business management level: fund guarantee (budget control and fund management), material and equipment management, infrastructure management, information construction management, etc.

Business operation level: living security and medical & health (consumption in prison, clothing, food, sanitation and epidemic prevention, disease medical treatment), material and equipment distribution, financial management, information service, logistics management, administrative file management, infrastructure construction, etc.

The comprehensive support business architecture is shown in Fig. 1.8.

Team building business is classified according to the functions of business decision-making level, business management level and business operation level. The corresponding business components (prison business functions) are:

Business decision-making level: performance appraisal system, quality evaluation system, role responsibility planning, human resource allocation plan, integrity risk prevention and control system, etc.

Business management level: performance appraisal management, quality evaluation management, anti-corruption risk prevention and control, discipline inspection and supervision management, personnel appointment and removal management, etc.

Business operation level: political and police personnel work, performance appraisal work, education and training work, supervision and audit work, etc.

The team building business structure is shown in Fig. 1.9.

Business synergies are classified according to functions at the business decision level, business management level, and business operations level, and the corresponding business components (prison business functions) are:

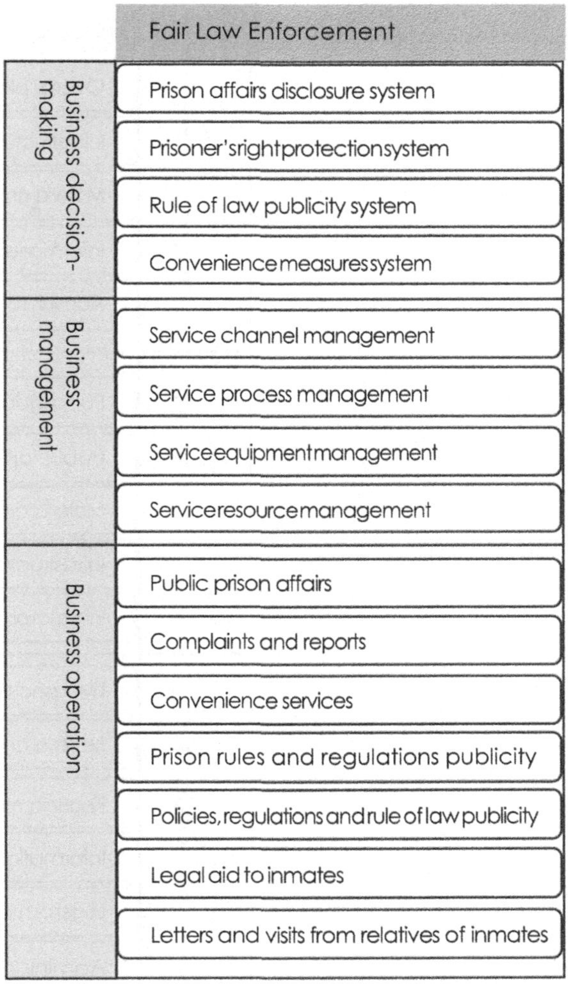


**Fig. 1.7** Fair law enforcement business architecture

Business decision-making level: business synergy mechanism requirements. Business management level: business collaborative process management.

Business operation level: external business coordination, supervisor and staff business coordination.

The business synergy business structure is shown in Fig. 1.10.

For comprehensive model definition and design ideas, the smart prison state of the art design business architecture blueprint is composed of the above business themes and business functional components, as shown in Fig. 1.11.

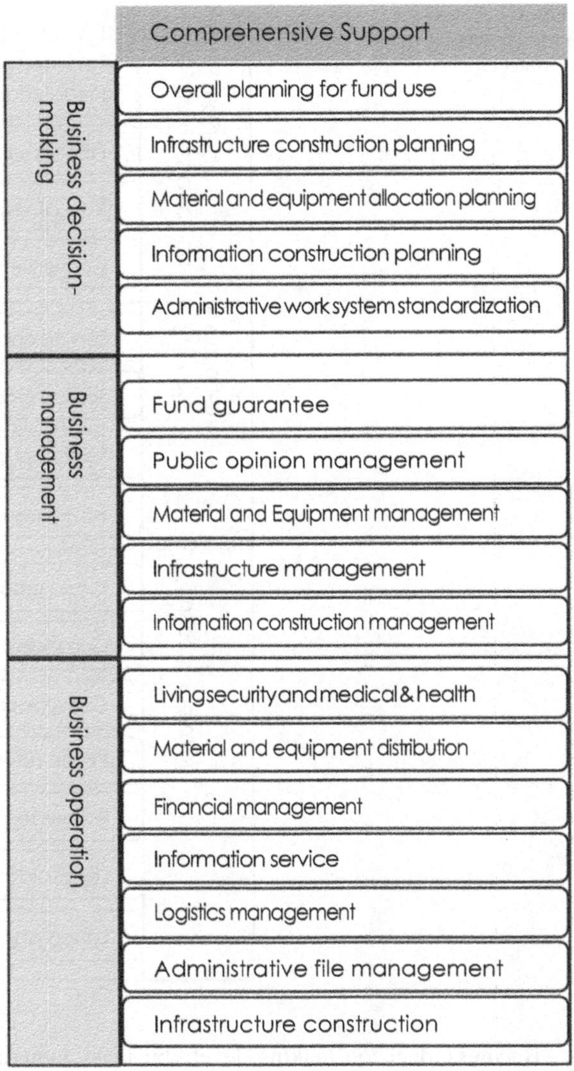


**Fig. 1.8** Comprehensive support business architecture

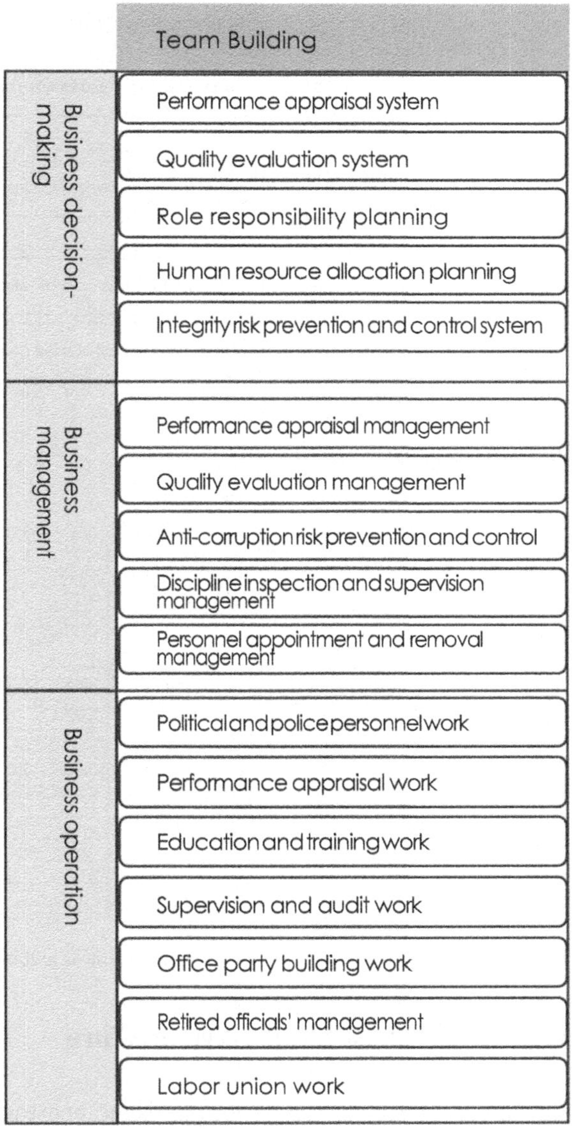


**Fig. 1.9** Team building type business structure

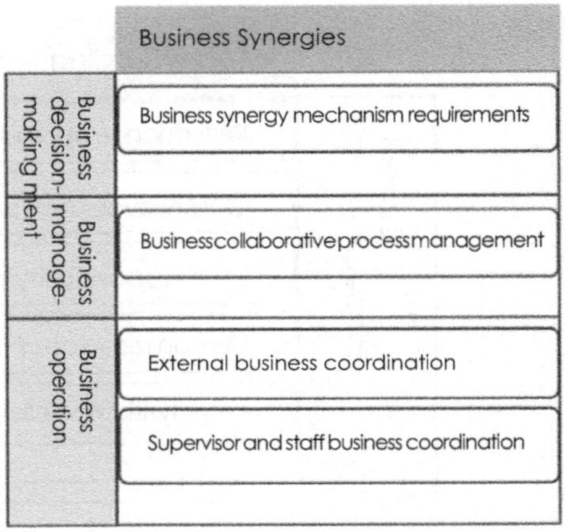


**Fig. 1.10** Business synergy structure

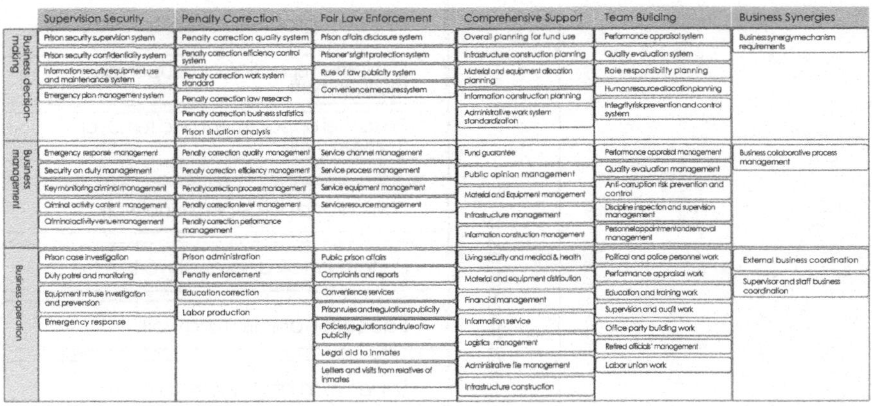


**Fig. 1.11** Smart prison state of the art design - business architecture blueprint

## 1.4 Smart Prison Data Architecture

Data is the core of the "state of the art design" construction of smart prisons, and it is also an important carrier to promote the integration of prison business and services.

The smart prison "state of the art design" data architecture consists mainly of the following:

(1) Data structure

The analysis and design of the prison data sharing and exchange system are carried out based on the classification of business types and manifestations, logical distribution of planning data, and the database that needs to be built for the "state of the art design" of the smart prison (Fig. 1.12).

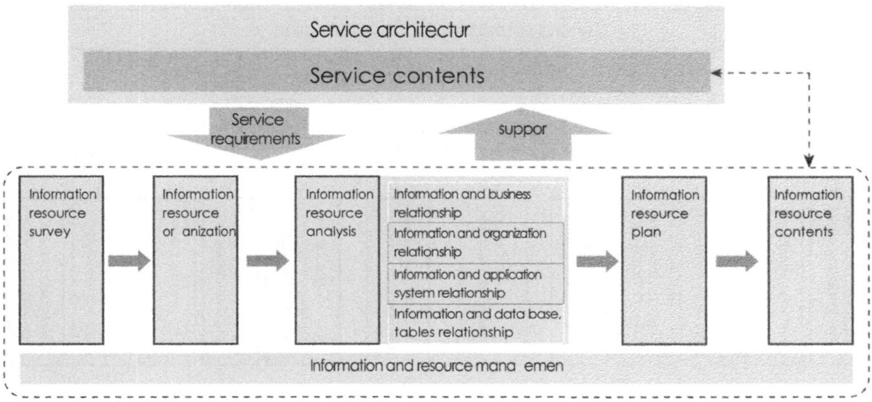


**Fig. 1.12** Data structure design

(2) Data modeling

Data modeling is an important part of the "state of the art design" construction of a smart prison. Based on its characteristics, operational data store (ODS) modeling, data warhouse (DW) modeling, and data mining (DM) modeling design are carried out.

(3) Data governance

Data governance is an effective means to ensure data quality and security. Through the "state of the art design" metadata management, data quality inspection and monitoring, data security management and data lifecycle management of smart prisons, the goal of achieving the "state of the art design" for smart prisons is realized.

(4) Data standards and specifications

Based on the characteristics of the "state of the art design" for a smart prison, relevant data processing guidance and management standards are established (Fig. 1.13).

(5) Information resource catalog

The information resource catalog is an important form of external display for data. Based on the analysis of data status, the information resource catalog of smart prison "state of the art design" is established and the relationship between information resources and service forms and databases is established to realize all aspects of prison data.

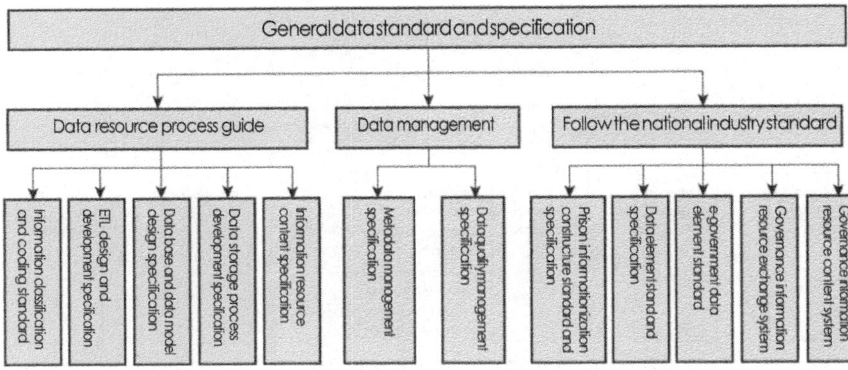


**Fig. 1.13** Data standards and specifications

## *1.4.1 Subject Domain*

The subject area is an abstract of the business model. Through the analysis of the prison business, a high-level view of the subject area of the smart prison data and the main relationship between the fields is obtained, as shown in Fig. 1.14.

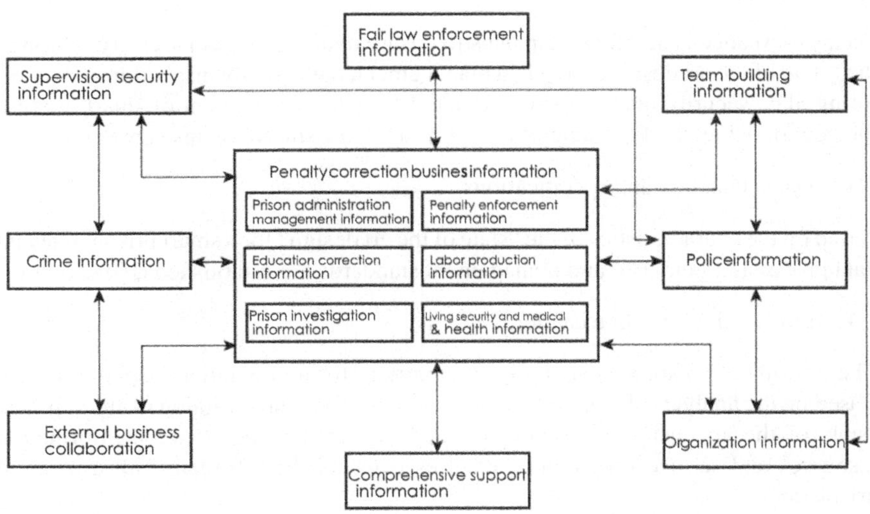


**Fig. 1.14** Smart prison data subject domain

### 1.4.2 Data Architecture

Data architecture is the information foundation for state of the art design and a data resource pool that guarantees state of the art design services. The overall data structure of the smart prison "state of the art design" is mainly composed of information resource catalogs, data architecture, data governance, data standards and specifications (Table 1.2).

The blueprint of the data architecture for the "state of the art design" of a smart prison is shown in Fig. 1.15.

**Table 1.2** The overall data structure

| Serial number | Name | | Description |
|---|---|---|---|
| 1 | The information resource directory | | The information resources are reviewed and a directory system is formed to facilitate access to relevant information resources, and the comprehensive integration of business service and data is realized through the directory system |
| 2 | The data stucture | Data classification | Classification is based on business entities and information representing classifications |
| | | Data logic distribution | The state of the art design database is divided into data sources (production libraries), raw data caches, Operational Data Store (ODS), Data Warehouse (DW), Data Mining (DM), semistructured/unstructured data vaults, collaborative shared exchange libraries, historical archives, and so on |
| | | Data modeling | Modeling and designing concept models, logical models, physical models, etc. for basic information libraries, integrated information libraries, analytical information libraries, etc |
| | | Data exchange and share | The state of the art design data migration process, data sharing exchange mechanism, data sharing exchange content, data sharing exchange and exchange technology are designed |

(continued)

**Table 1.2** (continued)

| Serial number | Name | Description |
|---|---|---|
| 3 | Data governance | This ensures state of the art design of a stable and efficient data management system, which mainly includes data life cycle management, metadata management, data quality management, and data security management |
| 4 | Data standards and specifications | Defines the data involved in the state of the art design and the data items, data attributes, data exchange and sharing interfaces, etc. Within the application, exchange and management of the data and provide support for data transmission and exchange, data application services and data management |

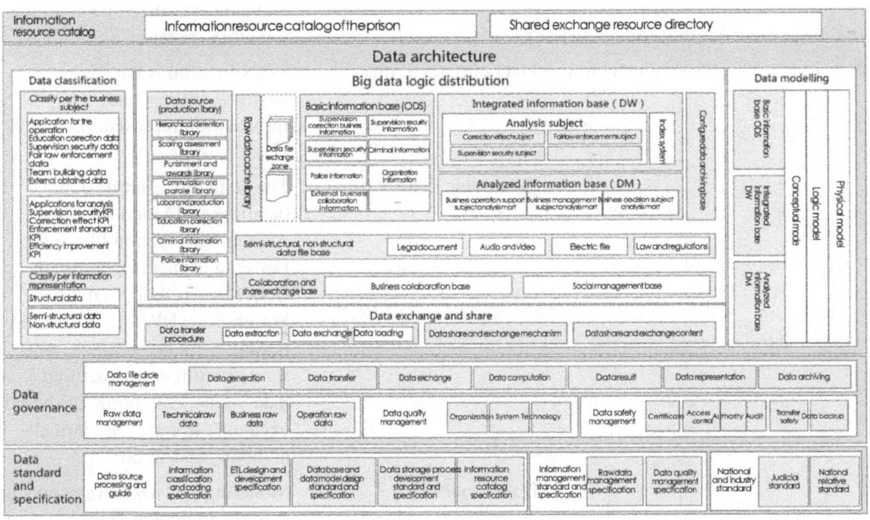


**Fig. 1.15** Smart prison state of the art design—data architecture blueprint

# 1.5 Smart Prison Application Architecture

## *1.5.1 Target Path*

Application architecture plays the role of unified planning. Connect the previous step with the next step by inheriting the business model and development direction upward and planning and guiding the positioning and function of each application system downward. This includes the application architecture blueprint, architecture standards and principles, system boundaries and definitions and contents such as the relationship between systems.

In smart prison application architecture, the word "smart" can be embodied in every application system through an auxiliary decision-making platform. Using the concept and technology of big data, data is extracted from the application system, and the path of "data→information→knowledge→intelligent" is realized. In practical applications, this link means "a large amount of fragmented data→fusing into not much information→transforming into very little knowledge" (Fig. 1.16).

Specifically, a series of key performance indicators (KPIs) for smart prisons can be used to measure achievement at four levels of goals: supervision security, law enforcement standards, correction quality and efficiency improvement. At the same time, one can establish a series of smart prison models to mine massive business data and to filter out big data models that are suitable for prison government goals and have high predictability. These models can help business management, business

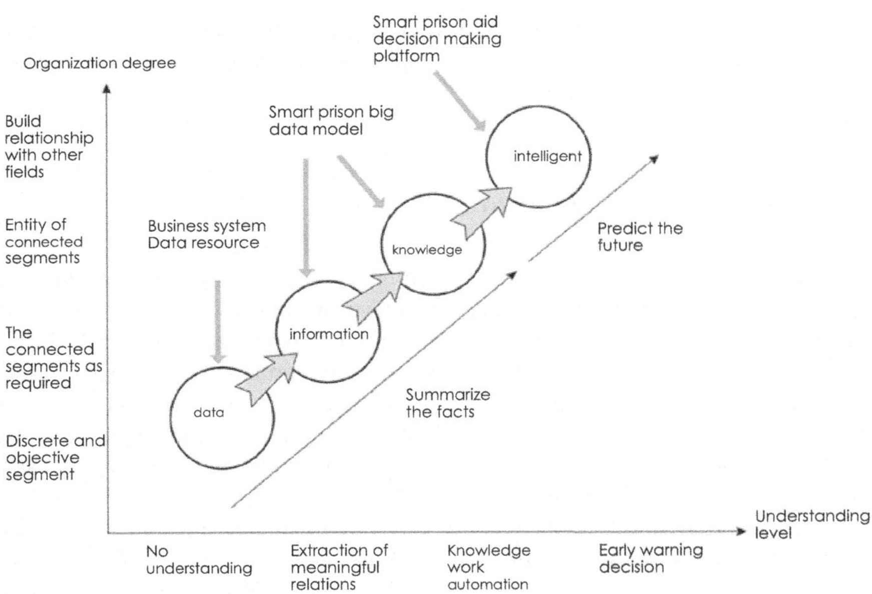


**Fig. 1.16** The implementation path of a "smart prison"

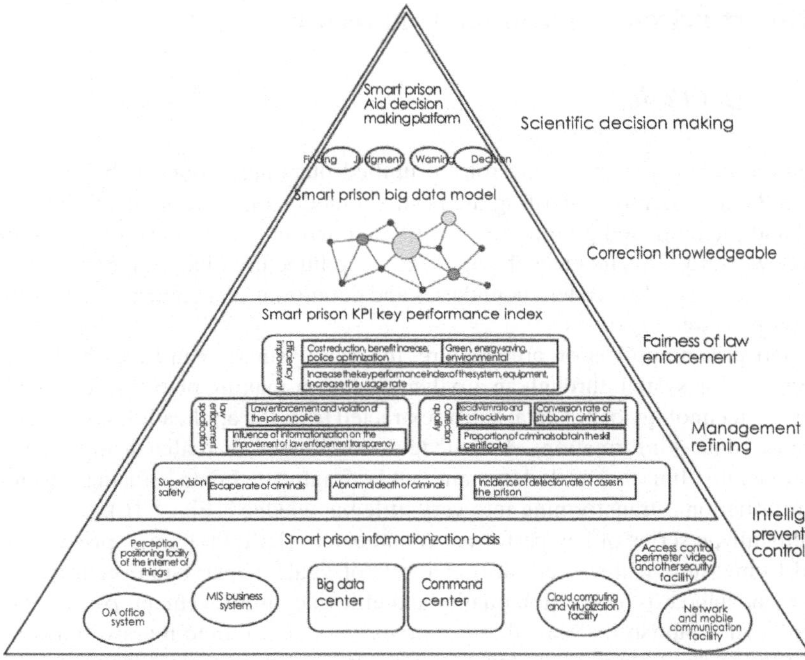


**Fig. 1.17** The pyramid structure of "smart prisons"

decision-making and business operations where future regulatory reforms will be carried out more effectively. The above framework can be described by the "smart prison" pyramid, as shown in Fig. 1.17.

## *1.5.2 Application Architecture*

The smart prison application architecture refers to the functional classification of business decision level, business management level and business operation level and adds public application levels. The corresponding smart work portals are the decision-making portal, the management portal, the business portal and the service portal. According to the actual situation of the prison application system and reflecting flat command thinking, the application themes can be divided into smart security, smart government, smart correction, smart law enforcement, decision-making early warning, prison affairs disclosure and judicial internet.

Smart Security: Information Security Integrated Linkage System, Emergency Command and Dispatch System, Internet of Things Positioning Supervision System.

Smart government affairs: OA system, mobile office platform, intranet website, message collaboration platform, police staff management system, prison construction support system.

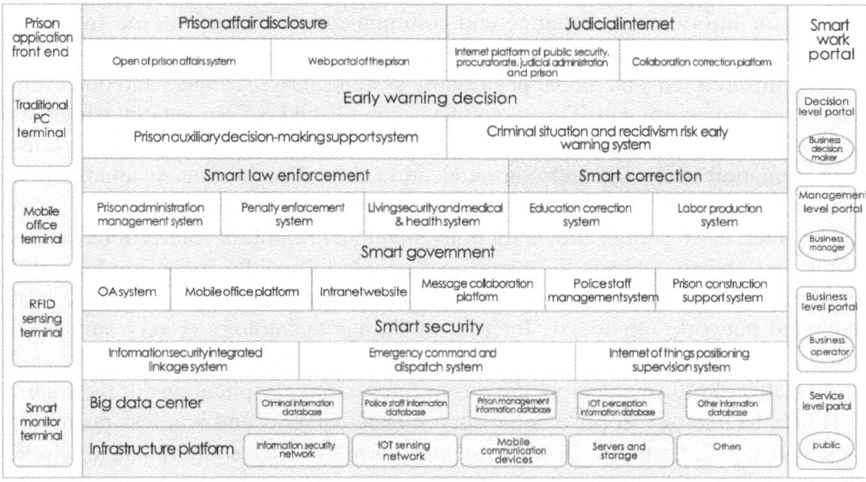


**Fig. 1.18** Smart prison state of the art design—application architecture blueprint

Smart correction category: education correction management system, labor production management system.

Smart law enforcement: prison administration management information system, penalty implementation management system, life security, medical and health systems.

Decision-making early warning: prison auxiliary decision-making support system, criminal situation and recidivism risk early warning system.

Prison affairs disclosure category: Open of prison affairs system, web portal of the prison.

Judicial internet category: internet platform of public security, procuratorate, judicial administration and prison, collaborative correction platform.

The blueprint of the "state-of-the-art design" application architecture for smart prison is shown in Fig. 1.18.

## 1.6 Smart Prison Technical Architecture

### *1.6.1 Key Technologies*

1. **Internet of Things**

The Internet of Things refers to the "Internet of Things Connected", which has two meanings. First, the core and foundation of the Internet of Things is still the Internet, which is a network that extends and expands on the basis of the worldwide website. Second, the user-side extension of the Internet of Things extends to any item and element for information exchange and communication. Therefore, the Internet of Things connects any item to the Internet through radio frequency identification RFID devices, infrared sensors, global positioning systems, laser scanners and other information sensing equipment. This would be accomplished per agreement using LAN, 3G/4G/5G, Wi-Fi, Bluetooth, ZigBee, UWB and other technologies that are used for information exchange and communication to realize intelligent identification, positioning, tracking, monitoring and management of a network.

The Internet of Things allows for convenient and seamless connections between people and objects, objects to objects and people to people. It can implement the real-time management and control of people, equipment and infrastructure in the integrated network. Obviously, Internet of Things technology is very suitable for prisons. In addition to the M2M (M: Man or Machine) linkage in regards to the application status of informatization, the most representative applications of the Internet of Things technology in prisons are surveillance cameras and wireless positioning for inmates, guard wireless positioning and alarm systems, shop tools and wireless positioning systems for dangerous goods, etc.

2. **Cloud Computing**

The core idea of cloud computing is to uniformly manage and schedule a large number of computing resources connected with the network to form a computing resource pool that can provide on-demand services to users. In a narrow sense, cloud computing refers to the delivery and usage mode of information technology infrastructure and refers to obtaining the required resources through the network in a way that is required and that is easily expandable. In a broad sense, cloud computing refers to the establishment of network server clusters by vendors to provide various types of customers with online software services, hardware rental, data storage, calculation analysis and other different types of services.

At the service level of cloud computing, the content provided by the state of the art design includes three levels: Infrastructure as a Service (IaaS), Platform as a Service (PaaS) and Software as a Service (SaaS). At the IaaS level, the state of the art design can provide infrastructure cloud services based on virtualization and provide the required infrastructure platform for various applications on demand. At the PaaS level, the state of the art design can provide application-supporting cloud services and a unified, on-demand application-supporting cloud service platform that supports the state of the art design of various services. At the SaaS level, the state of the art design can provide centralized information services and can provide personalized on-demand customization based on user permissions.

3. **Big Data**

Big data is another disruptive technological change in the IT industry following the Internet of Things and cloud computing. Big data refers to a collection of data that cannot be captured, managed and processed with conventional software tools within a certain time frame. It is massive with a high growth rate that requires new processing models to have stronger decision-making power, insight and discovery, process optimization capabilities and diversified information assets. In the "Big Data Era", written by Victor Meyer-Schoenberger and Kenneth Cukje, big data refers to the use of all data for analysis and processing instead of the shortcut of random analysis (sampling survey). The 5V characteristics of big data are volume, velocity, variety, value, and veracity.

After years of development, prison information construction has accumulated a large amount of basic data, including business data, audio and video materials, electronic files, legal documents, office documents and other data to which the development of the prison Internet of Things applications will bring geometric growth and diversified IOT perception data. These data contain many effective information. It is difficult to use conventional statistical analysis to address the complexity of big data. How then do we sort and reduce data noise? --by extracting and using information to provide for the decision-making and early warning of regulatory business. Proof will be the direction of development and exploration of smart prisons. How to effectively manage and apply this type of data will become an important technical requirement for the state of the art design of smart prisons.

### *1.6.2 Technical Architecture*

The technical architecture of the smart prison adopts the prison cloud management platform model. It supports the prison cloud management platform through four layers: information service layer, service support layer, information resource layer and basic resource layer, as well as the three major systems of standard and normative systems, safety guarantee systems and operation and maintenance guarantee systems. The four levels are:

Information service level (smart prison unified portal): basic information services, business collaboration services, prison affairs disclosure services, business decision-making services, application support services, business management services and business support services.

Service support level (prison information service bus/business collaborative sharing exchange platform): BI analysis engine, search engine, transmission and exchange, ETL, authority management, security audit, access control, data encryption, authentication, application middleware (routing/Load balancing, application pool management, message bus, cache management, IOT middleware).

Information resource level: data source area (production library), data exchange area, data integration area (information resource library, semistructured unstructured data file library), data archive area (historical archive), shared exchange area (collaborative shared exchange library), data management area (metadata database, data quality management library, ETL scheduling library, information resource catalog library).

Basic resource level: network system, host system, storage system, disaster recovery and backup system, security system, computer room environment, Internet of Things equipment.

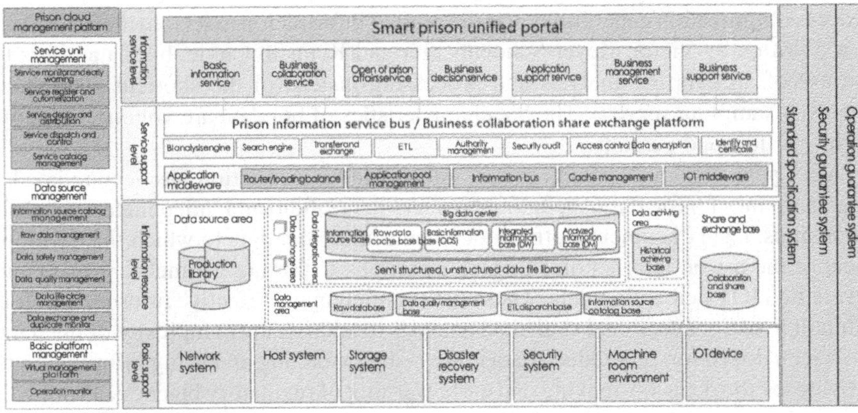


**Fig. 1.19** Smart prison state of the art design—blueprint for technical architecture

The blueprint of the "state of the art design" technical architecture of a smart prison is shown in Fig. 1.19.

## 1.7 Smart Prison Service Architecture

### *1.7.1 Service Reference Model*

In the "state of the art design" of a smart prison, as well as applying the TOGAF method, other related methods should also be considered. At present, the influence is relatively large, and the more widely used EA framework and methodology are also FEA (federal enterprise architecture). As the e-government state of the art architecture of the U.S. federal government, FEA is an important methodology worthy of reference. It is the basis for systematic analysis of the government's business processes, service capabilities, organizational components and technologies used. Its design purpose is to share government IT investment and achive the interconnection of government information resources (Fig. 1.20).

FEA consists of five reference models:

◎ Performance Reference Model (FEA-PRM): Provides a performance measurement framework for government agencies with general results and output indicators.

◎ Business Reference Model (FEA-BRM): Describes the internal operation of government agencies and the business process of providing services to citizens which constitutes the basic content of FEA.

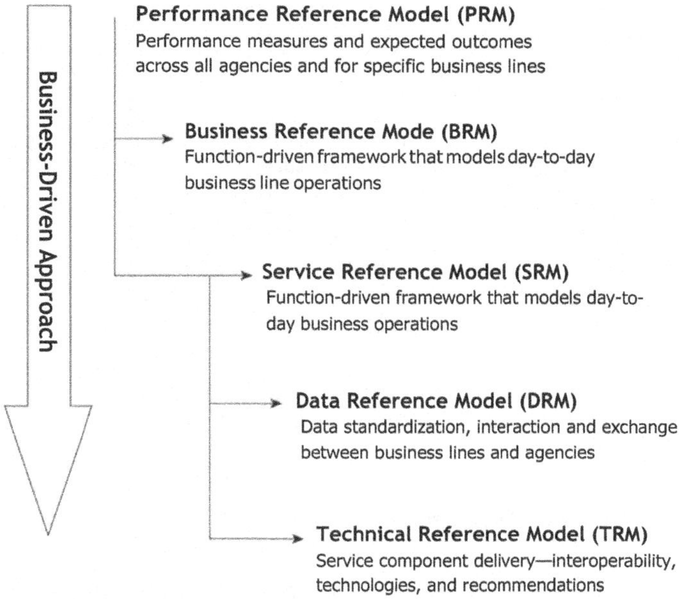


**Fig. 1.20** FEA reference mode. *Note* The original image is taken from https://obamawhitehouse.archives.gov/omb/e-gov/fea

◎ Service Component Reference Model (FEA-SRM): A component-based framework that can support application reuse, application system capabilities, components and business services. The component is a self-controllable business process or service that has been set up in advance.

◎ Data Reference Model (FEA-DRM): Describes the data and information that support the project plan and business flow operation process.

Technology Reference Model (FEA-TRM): A hierarchical technical architecture used to describe transmission service components and technical support methods to improve service performance.

Established by the scope of this book, the "state of the art design" is based on the four basic types of architecture covered by the original TOGAF combined with the FEA service component reference model in five aspects: business architecture, data architecture, application architecture, technical architecture, and service architecture. The overall structure of the smart prison design and the EA category adjustment is shown in Fig. 1.21.

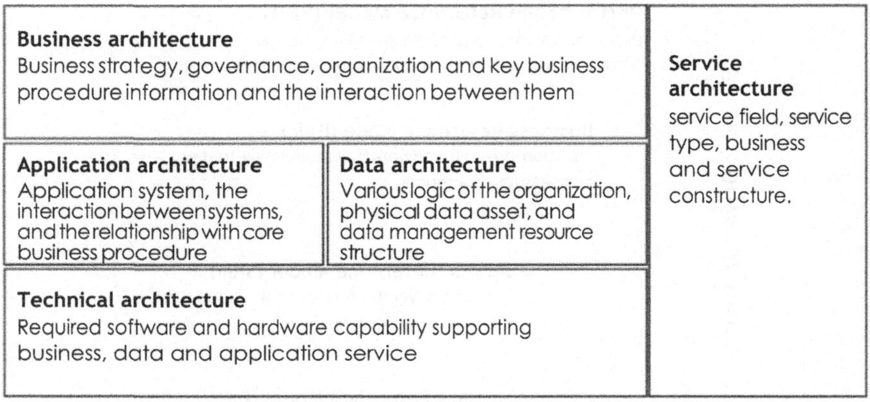


**Fig. 1.21** EA category 2

### *1.7.2 Service Architecture*

In the "state of the art design" of a smart prison, there are multiple relationships between various services and businesses. Under such a model, the ability to reuse and flexibly combine a single service is most important. On the one hand, with the help of the reusability of IT services, an IT service can provide auxiliary support for multiple types of businesses, reducing the development cost and implementation and maintenance costs. On the other hand, with the help of a flexible combination of IT services, business requirements are split into basic service units, and various service combinations are used to flexibly cater to changes in demand. This not only meets current needs but also enhances the risk resistance of the "state of the art design" system in the future (Fig. 1.22).

The service architecture is designed as follows:

Service targets: prison leaders, middle managers, grassroots officers, prisoners, the public, business systems, judicial internet, information office;

Service catalog classification: business support, business management, application support, business decision-making, prison affairs disclosure, business collaboration and basic information;

Service channels and methods: service interface, desktop computer, Internet, mobile terminal, mobile phone short message, email, telephone, instant message, management terminal, etc.

From the service architecture design to the realization of the information service layer of the technology platform, a service-oriented SOA architecture model is used. It combines different functional units of the application for services such as penalty correction, regulatory security, fair law enforcement, comprehensive protection, team building and business collaboration. Combined into service components these components are then standardized from loose couplings with elastic granularity

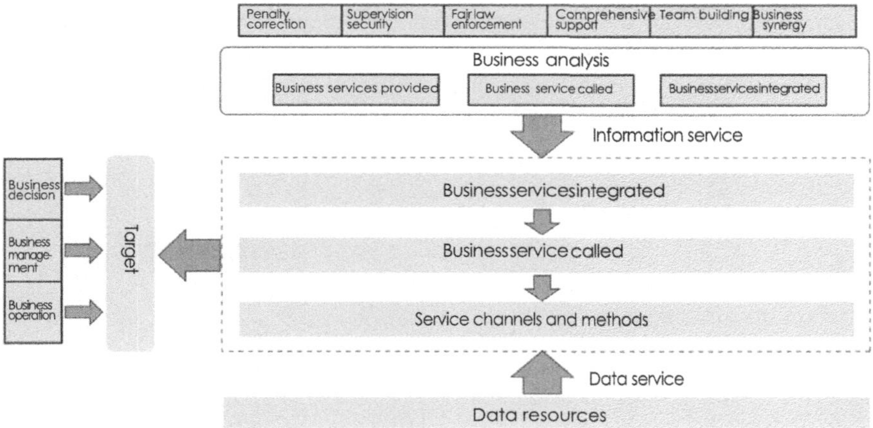


**Fig. 1.22** Service architecture design

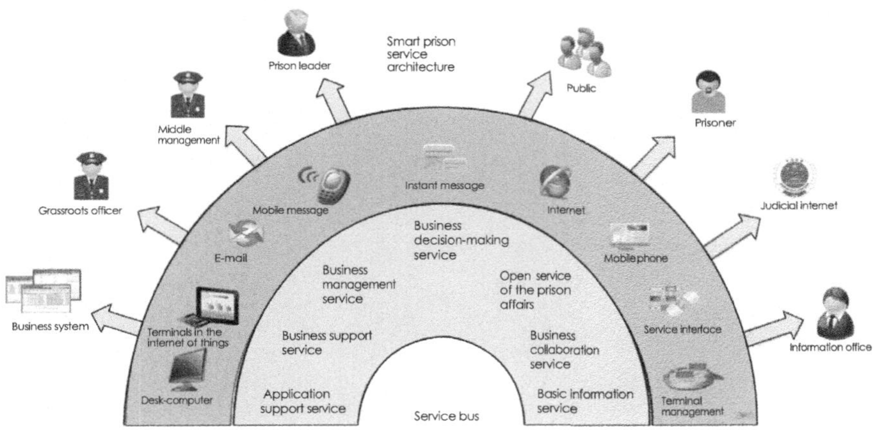


**Fig. 1.23** Smart prison state of the art design—service architecture blueprint

but on demand showing the characteristics of assembly to realize the comprehensive integration of business architecture, application architecture, data architecture, technical architecture and service architecture. Dynamic scheduling is used to serve different business needs, forming a fusion of information resources for the whole prison.

The blueprint for the "state of the art design" service architecture of a smart prisons is shown in Fig. 1.23.

# Chapter 2
# Evaluation Indicators for Smart Prison

## 2.1 Description of Evaluation Indicators

To scientifically evaluate smart prison construction, it is necessary to formulate a corresponding "Smart Prison Evaluation Indicator System".

The evaluation indicators for smart prisons adhere to the guiding principles that fully reflect the suggestions as follows:

1. The tasks and objectives of smart prisons adhere to the principle of rationality.
2. Evaluation indicators can be adjusted in time according to the actual development of the prison and adhere to the principle of operability to ensure that the evaluation is scientific and objective.
3. Fairness and justice adhere to the principle of standardization. Promote the establishment of a "standardization + informatization" two-wheel drive service system. "Lead informatization with standardization and promote standardization with informatization" to effectively support the modernization of prison management mechanisms and governance capabilities.

The evaluation indicators have the following characteristics. First, they can be collected. Historical and current data collection is scientific and convenient. Second, it is representative; they can more comprehensively reflect the overall development level of prison work. Third, comparisons between different prisons and different periods of the same prison can be made. Scientific comparisons can be made based on indicators. Fourth, it has scalability; the content of the indicator system can be increased, decreased, and modified according to actual development. Fifth, it provides guidance. The design of evaluation indicators should have important guiding values for the overall development of prison work.



P. Sun, *Smart Prisons*,
https://doi.org/10.1007/978-981-16-9657-2_2

## 2.2 Indicator System Design

The smart prison evaluation indicator system is divided into five categories (first-level indicators): concept planning, infrastructure, smart application, construction performance and support and assurance. The system includes 16 key evaluation elements (second-level indicators) and 44 important observation points. The measurement statistics of evaluation indicators take the form of hierarchical quantification and overall evaluation, assigning points according to the index evaluation level and then performing qualitative evaluation, turning subjective findings into objective data. The evaluation grade is divided into three levels: excellent, up to standard, and below standard. An excellent level is the standard that prisons are encouraged to achieve and the direction of the development of smart prisons. Up to the standard level is the basic requirement that smart prison construction should meet. The negative indicator is the implementation of a "one-vote veto" for supervision safety accidents, law enforcement violations and other matters. The evaluation indicator also encourages prisons to actively explore unique and innovative projects that suit them and that earn them extra points (Table 2.1).

**Table 2.1** Summary table of evaluation indicators for a smart prison

| Summary table of evaluation indicators for a smart prison | Key evaluation elements (Secondary indicators) | Important observation point |
|---|---|---|
| Idea planning | Construction concept | Leadership's awareness and attention |
| | | The understanding of the grassroots police |
| | Prison planning | Scientific and standardized planning |
| | | Infrastructure construction level |
| Infrastructure | Smart security network | Infrastructure construction level |
| | | Construction level of security engineering facilities |
| | | Application of Intelligent Technology |
| | Smart data center | Data center basic planning and architecture |
| | | Basic database construction level |
| | | Evidence preservation and disaster recovery backup |
| | | Data center operation management level |
| | | Intelligence degree of data center integration system |

(continued)

**Table 2.1** (continued)

| Summary table of evaluation indicators for a smart prison | Key evaluation elements (Secondary indicators) | Important observation point |
|---|---|---|
| | Smart command center | Command center infrastructure construction level |
| | | Command center operation management level |
| | | Emergency connectivity and emergency plan |
| | | Real-time early warning, collaborative command and electronic supervision capabilities |
| | | Public security, public prosecution and legal department interconnection |
| Smart application | Smart law enforcement | Business system specifications, functions and performance |
| | | Software integration and application level |
| | | Data and interface |
| | Smart correction | Business system specifications, functions and performance |
| | | Software integration and application level |
| | | Data and interface |
| | Smart government affairs | Business system specifications, functions and performance |
| | | Software integration and application level |
| | | Data and interface |
| Construction performance | Supervision security | Incidence rate and detection rate of cases in prison |
| | | Criminal escape rate |
| | | Unnatural deaths of criminals |
| | Correction quality | Reoffender conversion rate |
| | | Percentage of offenders obtaining cultural certificates and skills certificates |
| | | Disciplinary rate and risk of recidivism in prison |
| | | Recidivism rate |

(continued)

**Table 2.1** (continued)

| Summary table of evaluation indicators for a smart prison | Key evaluation elements (Secondary indicators) | Important observation point |
|---|---|---|
| | Law enforcement norms | Prison officers law enforcement violations |
| | | The influence of the information system on improving the transparency of law enforcement |
| | Performance improvement | Improve the key performance indicators of the system and equipment and increase its utilization rate |
| | | Cost reduction, efficiency improvement, police force optimization |
| | | Green, energy saving, environmental protection |
| Support guarantee | Normative system | Organizational Operation Management Standards |
| | | Information System Technical Specifications |
| | | Information security and operation and maintenance specifications |
| | Fund guarantee | Funding budget, funding source and expenditure, financial system and operation management |
| | Talent guarantee | Police professional structure and job configuration, personnel system and operation management |
| | Social support | Social participation and support |

## 2.3 Evaluation Index for Smart Prison—Concept Planning

Concept planning primarily includes two key evaluation elements: the construction concept and prison planning.

1. **Construction Concept**

Key observation points: 2 items, focusing on the concept of smart prison construction.

◎ Leadership's awareness and attention

Excellent standard: Advanced prison work concept, clear development ideas and goals and a strong sense of scientific management.

Up to standard: Attention to the study and research of advanced prison work concepts, clear development ideas and goals and scientific management awareness.

◎ Understanding of the grassroots police

Excellent standard: The grassroots police has an advanced understanding of the importance of smart prison construction.

Up to standard: The grassroots police has a sufficient understanding of the importance of smart prison construction.

2. **Prison Planning**

Important observation points: 2 items, focusing on whether the state of the art design, implementation steps, capital budget and other indicators of the plan are in harmony with the greater overall plan.

◎ The planning is scientific and standardized

Excellent standard: The prison construction and development plan is scientific and reasonable and conforms to unified technical standards.

Up to standard: A prison construction and development plan has been formulated that basically conforms to the standards.

◎ Implementation of the planning scheme

Excellent standard: The organization and implementation of the development plan and the management and supervision processes are clear. The results are significant.

Up to standard: The planning scheme can be implemented and progresses steadily.

## 2.4 Smart Prison Evaluation Index—Infrastructure

Infrastructure primarily includes three key evaluation elements: smart security network, smart data center and smart command center.

1. **Smart Security Network**

Important observation points: 3 items.

◎ Infrastructure construction level

Excellent standard: The prison network has complete hierarchical functions and effectively achieves security isolation. The prison private network bandwidth and performance are scientific and reasonable while meeting the standard requirements. It is forward-looking and exceeds the standard requirements. The basic network covers all terminal facilities inside the prison. The coverage rate exceeds the standard requirement. Up to standard: The prison network has fundamentally complete hierarchical functions and initially achieves security isolation. The bandwidth and performance of the prison private network meets the basic requirements. The basic network essentially covers all terminal facilities inside the prison, and the coverage rate meets the basic requirements.

◎ Construction level of security engineering facilities

Excellent standard: The functions and performance of access control, monitoring, intercom, perimeter, power grid, alarm and other systems exceed the standard requirements. The monitoring coverage of key locations within the prison and the proportion of high-definition cameras exceed the standard requirements.

Up to standard: Access control, monitoring, intercom, perimeter, power grid, alarm and other system functions and performance meet the basic requirements. The monitoring coverage and the proportion of high-definition cameras in key locations meet the basic requirements.

◎ Application of intelligent technology

Excellent standard: The investment proportion and construction level of the sensor network exceed the standard requirements and exceed the practical application level of new technologies such as the Internet of Things, cloud computing and big data.

Up to standard: Focused on research in the application of new technologies such as intelligence and posesses certain applications.

2. **Smart Data Center**

Important observation points: 5 items.

◎ Data center basic planning and architecture

Excellent standard: Data center layout planning, power planning, air conditioning and ventilation system planning, computer room equipment and IT infrastructure construction levels exceed the standard requirements.

Up to standard: Data center layout planning, power planning, air conditioning and ventilation system planning, computer room equipment and IT infrastructure meet the standard requirements.

◎ The level of basic database construction

Excellent standard: The electronic state of criminal information, police staff information, prison management information (prison administration management, penalty enforcement, education correction, labor production, hygiene, prison investigation, etc. information base) achieves 100% compliance.

Up to standard: The electronic rate of criminal information, police staff information, prison management information (prison administration, penalty enforcement, education correction, labor production, hygiene, prison investigation, etc. information database) meets the standard requirements.

◎ Evidence preservation and disaster recovery backup

Excellent standard: The prison exceeds the standard requirements in terms of evidence preservation, data storage and backup systems, backup data processing systems, backup network systems, backup infrastructure, technical support, operation and maintenance management, disaster recovery plans, etc.

Up to standard: The prison meets the basic requirements in terms of evidence preservation, data storage and backup systems, backup data processing systems, backup network systems, backup infrastructure, technical support, operation and maintenance management, disaster recovery plans, etc.

◎ Data center operation and management level

Excellent standard: The level of automated process management methods is excellent, and data center construction and application have achieved remarkable results.

Up to standard: The operation and management level meets the basic requirements, and the data center construction and application have achieved acceptable results.

◎ Degree of intelligence for the data center integration system

Excellent standard: Data center integration system (communication network system, office automation system, monitoring system, automatic fire alarm system, security system, personnel and asset positioning system, audio and video conference system, integrated wiring system, computer room engineering, etc.) exceeds the standard requirements.

Up to standard: The construction of the data center integrated system meets the standard requirements.

### 3. **Smart Command Center**

Important observation points: 5 items.

◎ Command center infrastructure construction level

Excellent standard: Command center graphic design, wiring design, ergonomics, large screen technology, acoustic/electromagnetic influence, environmental lighting, color coordination, seat design, security monitoring and alarm design, command room and computer room environment, etc. exceed the standards required.

Up to standard: The infrastructure construction of the command center meets the basic requirements.

◎ Operational management level of the command center

Excellent standard: The command center realizes automated operation, minimizes human intervention, accurate management tools and technologies, optimized processes and procedures and effective use of the physical environment.

Up to standard: The operation and management level of the command center meets the basic requirements.

◎ Emergency linkage and emergency plan

Excellent standard: Functions such as emergency duty, command and dispatch, security information integrated management, risk monitoring management, emergency resource management, emergency assessment management and other functions exceed the standard requirements. Emergency plans based on the regional characteristics and actual construction of the prison can be formulated and regular training can be conducted to strengthen the application of these plans via drill work.

Up to standard: Preliminary realization of emergency duty, command and dispatch, security information integrated management, risk and hidden danger monitoring and management, emergency resource management, emergency assessment management and other functions. An emergency plan has been formulated.

◎ Real-time early warning, collaborative command and electronic supervision capabilities

Excellent standard: Intelligent early warning and real-time alarm for information on the daily management of prisons, various emergencies and crimes to achieve rapid joint response; provincial bureau and prison bilevel command centers give full access to police coordination and electronic supervision of police law enforcement functions.

Up to standard: Able to undertake the daily management work of prison supervision, police supervision, remote command, etc., and achieve real-time early warning of various emergencies.

◎ Interconnection of Public Security, Procuratorate and Legal Affairs Department

Excellent standard: To realize the sharing and business coordination of security alarm information between the prison and the relevant departments in the prison. The sharing and business coordination of basic criminal information, criminal reform information and prison release personnel information with the Bureau of Justice. The public security sharing and business coordination of basic criminal information and identity verification information of visitors to the prison. Sharing and business coordination of basic criminal information, criminal reform information, criminal punishment implementation information with the procuratorate and criminal judgment information with the courts. Sharing and business coordination of commutation and parole application information and commutation and parole case ruling information.

Up to standard: The sharing of relevant information and business coordination between prisons and related departments within the prison, between the judicial bureau, between the public security, between the procuratorate and between the courts.

## 2.5 Evaluation Index of Smart Prison—Smart Application

Smart application mainly includes three key evaluation elements: smart law enforcement, smart correction, and smart government affairs.

1. **Smart Law Enforcement**

Important observation points: 3 items.

◎ Business system specifications, functions and performance

Excellent standard: Establish a unified and coordinated digital transformation standardization work mechanism in core business areas with clear powers and responsibilities. All core business specifications meet the requirements of the unified standard and the core business management software application coverage reaches more than 90%. Understands the intelligent assessment of criminal threat level and recidivism risk. Realize the online circulation, approval, supervision and inspection of criminals' administrative rewards and punishments, reception and release from prison, etc., as well as collaborative intelligence for daily law enforcement activities and personal hygiene management.

Up to standard: Establish management procedures and work standards for core business areas, core business specifications primarily meet the standard conditions and core business management software application coverage meets the basic requirements.

◎ Software integration and application level

Excellent standard: All software systems are integrated, the core business fully uses the intelligent analysis system to assist management and decision-making, the intelligent control and processing level of daily work is high, the software design promotion application process business department can play a leading role and the system is scalable, very flexible and upgradeable.

Up to standard: The software system is partially integrated, the core business initially uses intelligent analysis systems to assist management and decision-making, and daily work can be intelligently controlled and processed.

◎ Data and interface

Excellent standards: Prison administration management system and penalty enforcement management system, education correction management system, labor production management system, etc. provides a sharing interface of basic criminal information, criminal scoring and assessment information, criminal reward and punishment information, public health management system and prison affairs. The system is capable of sharing information on criminals' consumption in prison and

The data definition complies with the unified technical standards.

Up to standard: Data sharing is initially realized, and some systems provide connectivity.

2. **Smart Correction**

Important observation points: 3 items.

◎ Business system specifications, functions and performance

Excellent standard: Establish a unified and coordinated digital transformation standardization work mechanism in core business areas with clear powers and responsibilities. All core business specifications meet the requirements of unified standards, and the coverage of core business management software applications reaches in excess of 90%. Intelligent analysis and scientific evaluation of criminal education and correction requirements. The quality of criminal education and correction, the development of a resource library for criminal education and correction, the integration of high-quality social help and education resources and the application of intelligent collaboration.

Up to standard: Establish management procedures and work standards for core business areas, core business specifications essentially meet the standard requirements and core business management software application coverage meets the basic requirements.

◎ Software integration and application level

Excellent standard: All software systems are integrated, and the core business fully uses the intelligent analysis system to assist management and decision-making. The intelligent control and processing level of daily work is high. The business department plays a leading role in the software design promotion application process. The system is scalable, very flexible and upgradeable.

Up to standard: The software system is partially integrated, and the core business initially uses an intelligent analysis system to assist management and decision-making, and daily work can be intelligently controlled and processed.

◎ Data and interface

Excellent standard: The education correction management system, labor production management system, prison administration management system and penalty enforcement management system provide a correction information sharing interface. Data definition conforms to unified technical standards.

Up to standard: Data sharing is initially realized, and some systems provide connectivity.

3. **Smart Government Affairs**

Important observation points: 3 items.

◎ Business system specifications, functions and performance

Excellent standard: Establish a unified and coordinated digital transformation standardization work mechanism for core business areas with clear powers and responsibilities; All core business specifications meet the requirements of unified standards and the coverage of core business management software applications reaches in excess of 90%; Prison portals, prison public affairs and the electronic government affairs platform provide efficient circulation of official documents and intelligent collaboration; Use intelligent means to improve the daily management efficiency of the police and optimizes the allocation of the police force; Increases the supervision and assessment performance of the police and improves the efficiency of police inspection. In accordance with "green energy savings", the requirements of "environmental protection" provide the unified management and intelligent control of police technical equipment, buildings, canteens, hospitals, vehicles, warehouses, venues and other logistical support facilities, infrastructure and financial activities, resulting in reduced prison operating costs.

Up to standard: Establish management procedures and work standards for main business areas, core business specifications essentially meet the standard requirements and core business management software application coverage meets basic requirements.

◎ Software integration and application level

Excellent standard: All software systems are integrated, and the core business fully utilizes intelligent analysis systems to assist management and decision-making. The intelligent control and processing level of daily work is high. The business department plays a leading role in the software design promotion application process and the system is scalable, very flexible and upgradeable.

Up to standard: The software system is partially integrated, the core business initially uses an intelligent analysis system to assist management and decision-making and daily work can be intelligently controlled and processed.

◎ Data and interface

Excellent standard: The OA system and business system provide a document approval interface. Instant messaging systems and various business systems provide a message reminder interface. Email systems and various business systems provide an information push interface. The prison affairs disclosure system and prison administration management system can be graded. Provides an interface for assessment information sharing of the penalty enforcement system to determine the commutation and parole list. The interface of the police management system and other business systems to manage the sharing of police information, real-time data sharing, and data definitions that comply with unified technical standards.

Up to standard: Data sharing is initially realized and some systems provide connectivity.

## 2.6 Evaluation Index of Smart Prison—Construction Performance

Construction performance primarily includes 4 key evaluation elements: supervision security, correction quality, law enforcement norms, and efficiency improvement.

1. **Supervision Security**

Important observation points: 3 items.

◎ Incidence rate and detection rate of prison cases

Grade standard: The annual rate of crimes in prison is low. The investigation of in-prison cases is carried out effectively, and the annual detection rate is over 95%. There are no major or extraordinarily large cases in prison.

Note: This is a negative indicator.

◎ Inmate escape rate

Grade standard: No inmate escapes for three consecutive years. Note: This is a negative indicator.

◎ Abnormal death of criminals

Grade standard: Effectively provide psychological support for inmates, and no inmate suicide deaths occur. Provide suitable living quarters and sanitation. No food related deaths occur. Excel in safety production management, which results in no work-related inmate fatalities.

Note: This is a negative indicator.

2. **Correction Quality**

Important observation points: 4 items.

◎ Conversion rate of reoffenders

Excellent standard: The annual conversion rate of reoffenders meets the excellent standard.

Up to standard: The annual conversion rate of reoffenders meets the basic requirements.

◎ Percentage of offenders obtaining cultural certificates and skills certificates Excellent standard: Criminals who have actually served more than three years in prison should have at least one practical skill at the end of their sentence. For illiterate criminals who have the ability to receive education, the literacy rate will reach 100% at the end of their sentence. The proportion of criminals who obtained cultural certificates and technical grade certificates at the end of their sentence meets the standard of excellence.

Up to standard: Criminals who have actually served more than three years in prison should have at least one practical skill at the end of their sentence. For illiterate criminals who have the ability to receive education, the literacy rate will reach 100% at the end of their sentence. The proportion of the number of criminals who have obtained cultural certificates, skills, and grade certificates at the end of their sentence meets the basic requirements.

◎ Disciplinary rate and risk of recidivism in prison

Excellent standard: The accuracy of criminal risk evaluation analysis reaches the excellent standard.

Up to standard: The accuracy of criminal risk evaluation analysis meets the basic requirements.

◎ Recidivism rate

Excellent standard: The rate of recidivism within five years after the offender is released from prison meets the excellent standard.

Up to standard: The rate of reoffending within five years after the offenders are released from prison is not higher than the average rate within the province.

3. **Law Enforcement Norms**

Important observation points: 2 items.

◎ Violations of law enforcement by prison guards

Grade standard: The prison penalty enforcement system is sound. There are no cases of violations of laws or disciplinary actions in the process of commutation, parole and temporary performance outside of the prison. No cases of prison police being charged criminally. There were no cases of police abuse of criminals by way of physical punishment, corruption or bribery. No cases that have been held criminally responsible for dereliction of duty.

Note: This is a negative indicator.

◎ The impact of information systems on improving law enforcement transparency

Excellent standard: The information-based law enforcement system can minimize ambiguities and provide greater transparency, which promotes the fair and standard enforcement practices of the police. The information-based law enforcement system plays a role in regulating the law enforcement behavior of the police.

4. **Performance Improvement**

Important observation points: 3 items.

◎ Improve the key performance indicators of the system and equipment and increase its utilization rate

Excellent standard: The key performance indicators of prison systems and equipment have been greatly improved in reliability (time between failures), availability (average repair time) and maintainability. The utilization rate of systems and equipment has been greatly increased, exceeding the standard requirement.

Up to standard: The key performance indicators of prison systems and equipment have improved in reliability (time between failures), availability (average repair time), and maintainability. The utilization rate of systems and equipment has increased to meet basic requirements.

◎ Cost reduction, benefit improvement, police force optimization

Excellent standard: The complexity of informatization, investment and operation costs are effectively reduced. Rapid repetitive construction can be realized. Police force allocation is optimized, and police force shortages are effectively alleviated. Overall benefits are greatly optimized.

Up to standard: Investment and operating costs have been reduced. Police force allocation and overall benefits have been optimized.

◎ Green energy saving, environmental protection

Excellent standard: Through index testing, the prison's comprehensive energy consumption, cooling and heating demand and carbon emissions have been effectively reduced.

Up to standard: Through index testing, the prison's comprehensive energy consumption, cooling and heating requirements and carbon emissions have been somewhat reduced.

## 2.7 Evaluation Indicators for Smart Prison—Support and Guarantee

Support guarantees primarily include 4 key evaluation elements: standard systems, financial guarantee, talent guarantee, and social support.

1. **Standardized System**

Important observation points: 3 items.

◎ Organizational operation management norms

Excellent standard: According to the needs of prison smart applications, standardize the organization and management system, optimize and reconstruct the prison work process, clarify the job responsibilities of the prison command center, data center and other related departments and explore the establishment of professional job sequences in the field of information management and smart applications.

Up to standard: The organization and management system basically meets the requirements of the specification, and the job responsibilities of the relevant departments are essentially clear.

◎ Information system technical norms

Excellent standard: Establish information technology specifications covering various prison business data exchanges and network constructions. This includes basic data standards, network platform construction standards, business platform construction standards and data exchange standards.

Up to standard: Primarily establish information technology specifications that comply with prison business data exchange and network construction.

◎ Information security and operation and maintenance specifications

Excellent standard: In accordance with relevant national laws, regulations, standards and policy requirements, organize the establishment and implementation of a complete information security system and operation and maintenance system to ensure the security of prison networks and information systems.

Up to standard: The information security system and operation and maintenance system have been initially established, and the prison network and information system security environment has gradually improved.

2. **Funding Guarantee**

Important observation points: 1 item.

◎ Funding budget, funding sources and expenditures, financial system and operation management

Excellent standard: Construction funds are included in the fiscal budget and a dynamic guarantee mechanism for project construction and operation and maintenance has been established. Multiparty fundraising channels have been established and the ability to raise funds for project construction is strong. A fund use management system has been established, strict supervision and evaluation and good fund use performance.

Up to standard: Emphasize the investment in construction funds and initially establish a dynamic guarantee mechanism for project construction, operation and maintenance. Provides a capital use management system.

3. **Talent Guarantee**

Important observation points: 1 item.

◎ Police professional structure and job description, personnel system and operation management

Excellent standard: Establish a professional talent training and cultivation system by layering and classification; form a good talent growth incentive mechanism, promote the professional construction of the police force and provide a high degree of job satisfaction within the police.

Up to standard: The professional talent training system was initially established, and the professionalization of the police force was actively promoted. The interviews with the police are essentially satisfactory.

4. **Social Support**

Important observation points: 1 item.

◎ Social participation and support

Excellent standard: Widely utilize social resources; participation of enterprises, universities and research institutes. The level of social support is excellent.

Up to standard: The level of social participation and support is satisfactory.

# Part II
# Intelligent Security Protection

# Chapter 3
# Prison IOT

The Internet of Things in prison refers to technology installed in the prison, such as the RFID radio frequency identification of objects that is worn by the inmates and police, various sensors, global positioning systems, laser scanners and other information-sensing equipment used to collect prison-related information according to protocols. Through these connections, communication and dialog (e.g. M2M) between man and object and between object and object is realized, thus providing information to achieve a network of identification, positioning, tracking, and monitoring leading to refined management.

Based on the maturity and scale application of sensing technology, the Internet of Things in prison provides comprehensive and effective data support for the implementation of big data applications in prisons. From the perspective of infrastructure construction, the smart prison is a prison utilizing the internet of things.

## 3.1 System Architecture of the Internet of Things

The Internet of Things is a network that, through information sensing devices according to protocols, realizes the comprehensive interconnection between man and man, man and object, or object and object. Its main characteristics are to gain various information regarding the physical world through information sensing equipment and other means, to transmit and interact with this information through combining the internet, communication networks and other networks, and to analyze and process the information through the use of smart computing technology to improve perceptions of the physical world and to help realize decisions and controls resulting from this intelligence. The system architecture of the internet of things can be divided into three layers: a perception layer, network layer, and application layer, as shown in Fig. 3.1.

 
P. Sun, *Smart Prisons*,
https://doi.org/10.1007/978-981-16-9657-2_3

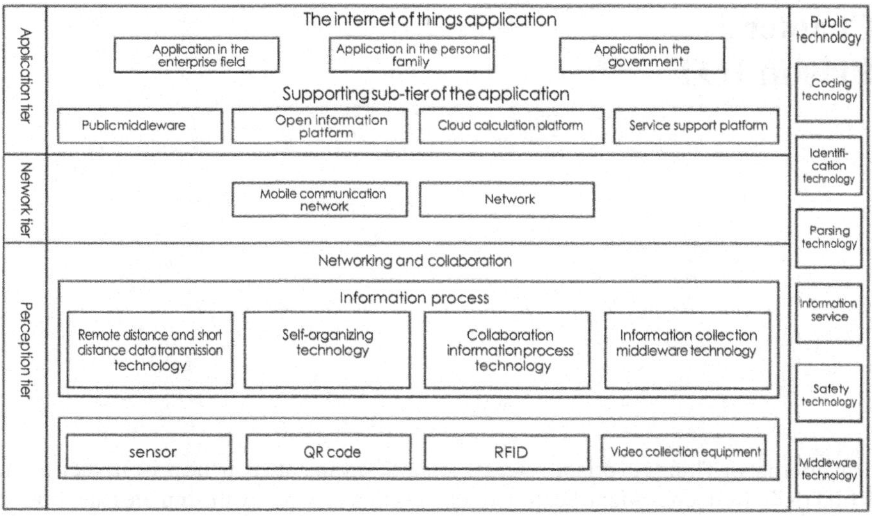


**Fig. 3.1** System architecture diagram of the Internet of Things

The perception layer is composed of equipment with perception and identification functions, including vibration sensors, carbon dioxide concentration sensors, temperature sensors, humidity sensors, QR code tags, RFID tags and readers, video cameras, GPS and other sensing terminals. This perception layer is equivalent to the nerve endings of human eyes, ears, nose, throat and skin. It is the source and function of the internet to identify objects and collect information.

The network layer is composed of various self-organized networks, the internet, wired and wireless communication networks, network management systems and cloud computing platforms, which is equivalent to the human nerve center and the brain and responsible for transmitting and processing information obtained by the perception layer.

The application layer is the interface between the internet of things and the user (including the human, the organization and other systems), which is combined with industry demand, to realize the smart application of the Internet of Things. The industrial characteristics of the internet of things is mainly reflected in its application area. Currently, smart logistics, power grid management, green agriculture, industrial monitoring, public safety, urban management, telemedicine, smart homes, smart transportation, environmental monitoring and other industries have all experimented with the application of the Internet of Things.

## 3.2 Perception and Recognition of the Internet of Things

The perception layer is the final stage of the Internet of Things, and the function of specific perception is achieved by sensor nodes. The sensor nodes consist of four parts: sensor module, processor module, wireless communication module, and energy supply module.

The sensor module is responsible for monitoring the collection of information and conversion of data; the processor module is responsible for controlling the operation of all sensor nodes, storing and processing the collected data and the data that is sent from other nodes; the wireless communication module is responsible for wireless communication with other sensor nodes, exchanging control messages, and sending and receiving data collected by the sensor; and the energy supply module provides the energy required for the operation of the sensor nodes, normally by using microbatteries.

Automatic identification is an important method to collect and read the data and input them into the computer automatically. The collected data is the information of the objects to be identified, i.e., the objects are about individuals, animals, goods, etc. In the past two or three decades, automatic identification technology has rapidly developed all around the world and is quickly being applied to commercial departments, production enterprises and material circulation fields. A high-tech discipline that covers bar code identification technology, radio frequency identification technology, biometric identification technology, image recognition technology, magnetic stripe recognition technology, etc., integrating communication network technology, computers, light and electricity has been preliminarily formed.

The classic perception technology of the Internet of Things has approximately the following types:

1. **Visual perception technology**

Visual perception technology is of a noncontact type, such as the integration of video camera equipment and other technologies. The collected image data is transmitted to a back-end video analyzing server through the front-end image sensing device and stored. The video server analyzes the collected video image information and separates the person or the object in the video from the background for analysis and tracking. According to preset safety rules, compared with the behavior pattern of the tracked targets, if abnormal conditions or violations appear, one can call the police and notify them immediately while recording or displaying the relative information.

In fact, achieving full coverage of prison law enforcement monitoring is still a serious challenge to the police force in prisons that are under traditional manual management. Traditional methods cannot warn the police in advance, so video intelligent analysis technology is a progressive development. The latest video intelligent analysis technology can even find a person who is highly nervous in a crowd, which provides the police a tool to identify a dangerous attacker effectively. In addition to automatically recognizing human faces, many cameras can also coordinate with each other to achieve tracking relay of video monitoring. By combining the collected video images with data collected by other sensors in the internet of things in the prison, it is possible to recognize dangerous behavior effectively, which saves time for prison police and improves prison supervision, safety, and law enforcement.

2. **Auditory perception technology**

With the development of computers and phonetics, using sensors to replace human ears is being achieved. By recognizing the speaker with the technology of speech processing and recognition, it is possible for the system to correctly understand some simple sentences. From the perspective of application, the system of recognizing sound can be divided into two categories:

(1) Speaker recognition system

The purpose of the speaker recognition system is to distinguish whether the received sound is the sound of the person who is identified in advance and whether the sound is the sound of some person in a group specified in advance.

(2) Semantic recognition system

The semantic recognition system is based on the basic principles of pattern recognition. When the system is operating, the received voice signal is compared with the standard pattern to recognize the meaning of a voice signal.

Acoustic sensors are usually microphones or hydrophones that convert sound waves from the sound source generated by the vibration of the air or the water. This is an important electroacoustic device. The technical characteristics of the microphone mainly include sensitivity, frequency response, directivity, and dynamic range.

The sound sensors installed in the prison could play an important role in the early warning of emergencies by noise. Meanwhile, they could be a new, reliable technical method for detection in prisons.

3. **Environmental perceiving technology including temperature and humidity, etc.**

Temperature and humidity sensors can be installed in important locations where the requirements of temperature are stricter to monitor real-time temperature changes. When incidents occur, the system will call the police in time. For example, by a prisoner wearing an electronic wristband with a built-in micro temperature, pulse or blood pressure sensor, you can evaluate the person's vital signs in real time and give timely warnings which can effectively monitor and predict emergencies, such as the self-harm, suicide, sudden death of a prisoner, and so on.

4. **Displacement perceiving technology**

Displacement perceiving technology can measure linear and angular displacement. The displacement sensor could be installed in weapons and dangerous working tools, important assets and fixed facilities in prison, such as alarms, video cameras, firearms, vehicles, keys, dangerous tools, etc. The various attributes and locations of these could be sent to the asset management system by a wireless sensing network and computer network. When the sensors of these fixed important assets and facilities indicate movement, an early warning in real time could be issued.

5. **Acceleration perceiving technology**

Linear acceleration refers to the acceleration at the center of gravity of an object moving along its motion track. It is a basic physical quantity that characterizes the nature of the movement in space. The state of motion of the object can be measured by measuring acceleration.

By integrating the acceleration sensor into an electronic wristband worn by prisoners, in special application periods, such as during sleep, once there is continuous acceleration perceiving signals, which may indicate violent motion, the video could be connected. People could use the closest video camera to effectively avoid the occurrence of some incidents, such as fighting among prisoners or attacking the police.

6. **Attitude perceiving technology**

The attitudes of a moving body in the objective world include the angle, angular velocity and angular acceleration of the moving body, which can be obtained and perceived through gyroscope technology. The attitudes (heading, pitch, roll) of the moving body could be measured through the fixed axis and precession of the gyro. The angular motion of the moving body could be measured precisely, which could achieve a stable inertial platform through the inertial coordination system composed within the gyro.

By integrating a micro gyroscope into electronic wristbands and police tags worn by prisoners, the ability to monitor the behavior of criminals and the police could provide a real-time early warning for possible violent behaviors of dangerous criminals.

## 3.3 Applications of the Internet of the Things

### *3.3.1 Architecture of the Internet of Things in Prisons*

Currently, security technology is evolving rapidly and new technology is emerging. Prisons are places with extremely high-level requirements for security. It is a misunderstanding to pursue these technologies excessively. The prison should not become a testing ground for so-called new technology, especially when lacking scientific and unified state of the art design which is a major challenge for balancing security and new technologies.

Prisons have their government goals, including security supervision, education and correction, and fair law enforcement. To evaluate the applicability of a security technology to prisons, it is necessary to focus on the above objectives. The evaluation standard is to see whether it helps to improve the ability of safety supervision, whether it helps to enhance scientific education and correction, whether it helps regulate the management of law enforcement, whether it helps to improve working performance and efficiency, and whether it helps to enhance the quality of policing (police with science and technology).

The security construction of smart prisons focusing on the Internet of Things is currently a concern. As an important place to ensure social public security because of its relatively enclosed space with dense occupancy, complex composition, and high level of protection, there is an urgent need for intelligent management. The Internet of Things allows seamless connection between man and object, object and object, man and man. Through information sensing equipment, such as radio frequency identification, wireless positioning systems, video cameras, and access control perimeters, according to protocols, connecting the object with the internet could implement real-time management and control for personnel, equipment and infrastructure in the integrating network. Obviously, the technology of the Internet of Things is very suitable for the current status of the application of the informationization in prison.

The overall architecture of the Internet of Things in prison is shown in Fig. 3.2.

The barrier between civil defense, material defense, technical defense and joint defense is broken down through the technology of the internet of things in prisons. It realizes the real joint linkage, control and protection for civil defense, material defense, technical defense, and joint defense. It improves overall security support capabilities in prisons.

The flow and the change of the persons, objects, and environments that prison management supervises is controlled in real time through the technology of the internet of things in prisons. This should ensure that the process is safe and that law enforcement management behavior is standardized to refine prison management.

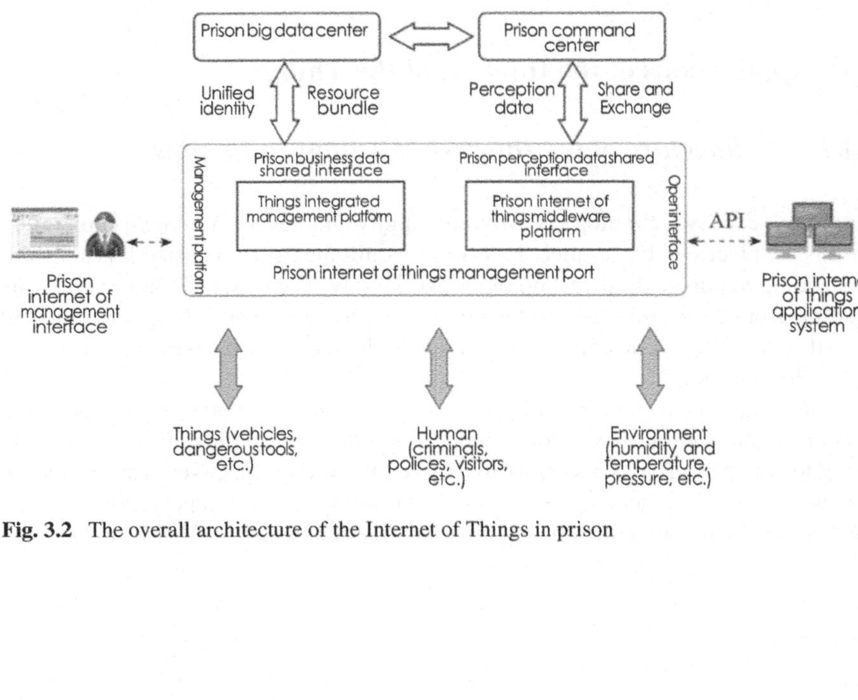


**Fig. 3.2** The overall architecture of the Internet of Things in prison

The vital signs and abnormal behavior of criminals can be perceived in real time through the technology of the internet of things in prisons, which can provide key support for the establishment of psychological evaluations and risk indicators, along with better implementation of individulized education and correction.

### *3.3.2 Three Stages of the Development of the Internet of Things in Prisons*

Generally, the Internet of Things is a network of connections between objects. There are three stages for the application of the Internet of Things that can be explained in M2M (M: man or machine).

The first stage is "machine to machine", which mainly solves problems in joint linkages between objects. Currently, the informationization of prisons in China is at this stage. As the central hub of prison management, the establishment of the command center is the best practice at this stage.

The second stage is "machine to man" or "man to machine", which mainly solves problems of communication between object and man; this stage could also be called the semi-intelligence stage. At this stage, the focus is solving problems of informatization and tagging of people (criminals, police, visitors, etc.). Criminal electronic wristbands, police electronic badges, etc., have played a role in converting the individual into an information node to access a network.

The third stage is "man to man", that is, communications between networking objects, where objects and humans are close to or reach human-like levels, and that stage could be called the intelligent stage.

The original meaning of M2M (machine to machine, communications between machine and machine) is that the applications and services are focused on the intelligent interaction of machine terminals. It is a kind of interacting communication between machine and machine that gives machine intelligence. From the perspective of infrastructure, a smart prison is a prison with the Internet of Things. Compared with the understanding of traditional informationizing construction of the prison-MCP (MIS + camera + perimeter, information system + camera + perimeter protection system), the characteristics of the Internet of Things, including active security, real-time linkage, precise control, etc., can provide an opportunity for the reform of the supervision pattern of the prison.

Although it is predictable that the development of the Internet of Things in prisons is coming, there is no doubt that application is still at a primary stage. As a core component, the status of the sensing network in the internet infrastructure is currently still limited. In the application field of people, object, environment sensing and perceiving, apart from the applications of pilot wireless positioning, only the applications of video monitoring cameras and pickups are mature, and there is still a long way to go for the development of these two in the field of intelligence (intelligent voice/video analysis). Similarly, it is inevitable that there are some problems during the development of the Internet of Things, including technology, standards or industry application, and privacy, safety, etc.

### 3.3.3 Middleware of the Internet of Things in Prisons

With the gradual start of pilot construction of the sensing network in prisons, especially with the continuous increase in the pilot RFID application of the prison sensor network based on wireless positioning, problems integrating various heterogeneous devices in prisons are increasingly prominent. Due to the lack of related standards of the Internet of Things in prisons and the differences of the products between the manufacturers, each pilot prison needs to develop a new linkage application, which lacks software reusability and is hard to promote, especially during the construction of the command center. Due to the lack of the related standards, the data that is generated from the sensor in the current system of the internet of things are often "kidnapped" by each system. Therefore, it is urgent to solve the problems of construction standards in the construction of the Internet of Things in prisons to deeply integrate, share and link various systems and various business systems.

The middleware of the Internet of Things in prisons (Prison IOT Middleware, PIM) is designed to address this problem. Based on service-oriented software architecture, using the standard service interface, the middleware could provide service to top application systems. It could conduct unified personnel positioning, video monitoring, access and perimeter control across heterogeneous systems. The architecture of this middleware platform is shown in Fig. 3.3.

Taking specific wireless positioning in prisons as an example, through the construction of the middleware platform of wireless positioning in prisons (the subplatform of the Internet of Things in prisons), the pilot application of wireless positioning could be compatible with different manufacturer and technology selections. The platform system can back up, filter, integrate, and release the collected personnel positioning, video monitoring, access control and perimeter control data, standardize various information formats and protocols, and provide valuable data to top application or data management systems. Meanwhile, it shields the differences between various RFID readers to realize transparent integration between various parts, aids in the development of relative application systems, and plays an intermediary role between the RFID reader and application system.

The middleware system of wireless positioning in prisons, as shown in Fig. 3.4.

(1) **Message routing module**

According to the device information of the given target, this module mainly gains the device marking of specific systems and the address information required by the transmission information and is responsible for delivering the message.

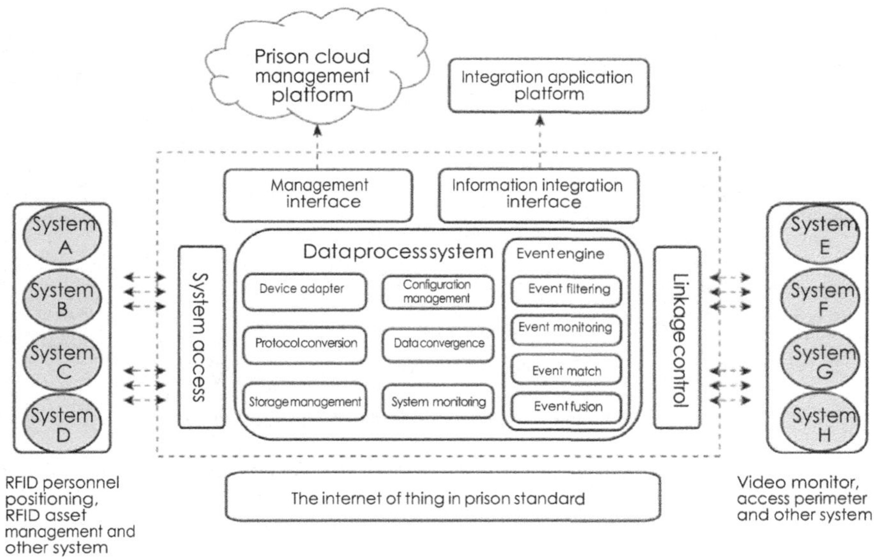


**Fig. 3.3** The architecture of the middleware platform of the Internet of Things in prisons

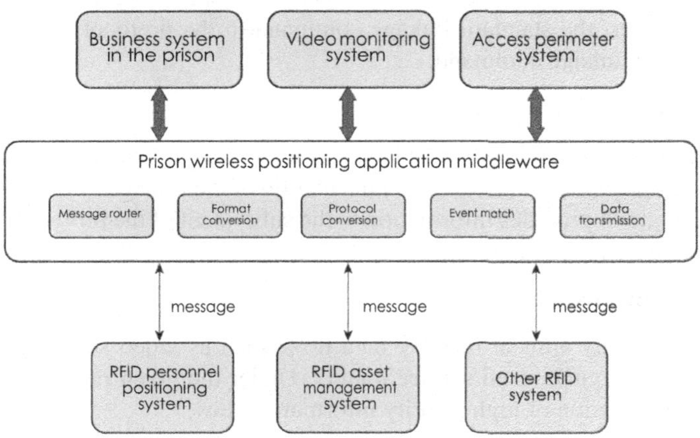


**Fig. 3.4** The middleware system of wireless positioning in prisons

## (2) Format conversion module

This module mainly includes information, format conversion and reverse conversion in the protocol, including the conversion of specific device positioning information into standard position information.

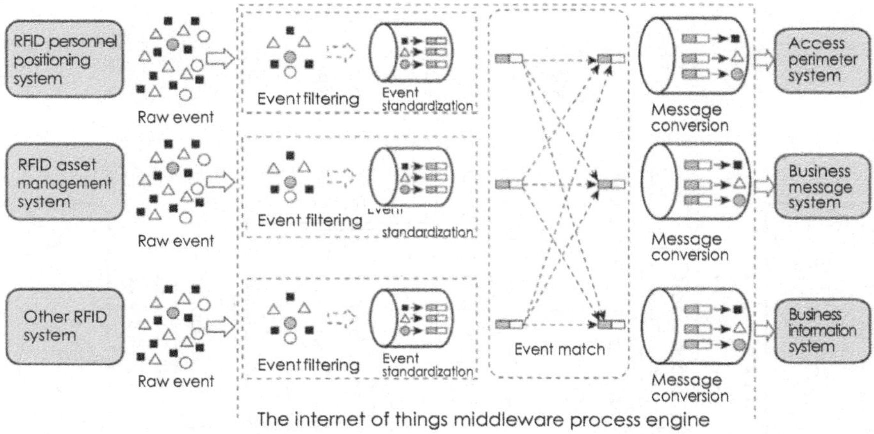


**Fig. 3.5** Event processing engine of the middleware of the wireless positioning application in prisons

(3) **Protocol conversion module**

This module mainly accomplishes two goals: one is to convert the original event information of various heterogeneous devices into the standard event format, and the other is to convert the standard linkage command in the protocol into the specific device-oriented linkage command.

(4) **Event matching module**

According to the configured specific business logic, this module mainly determines the target equipment and other related information of the linkage information generated by the logic. It provides information to the subsystem of the protocol. The event matching module is shown in Fig. 3.5.

(5) **Data transmission module**

This module mainly aims at massive data flow, such as video streams, to provide transmission with guaranteed service quality. Under the given resource, it ensures the stable transmission of high priority information flow.

## 3.3.4 *Internet of Things Security in Prisons*

In the application of the Internet of Things in prisons, without information security protection measures, not only can personal information be easily leaked, but it is also possible for hackers to invade and gain administrative authority over the Internet of Things system control. The information security issues faced by the Internet of Things in prisons mainly include perceived terminal security, perceived network security, transmission network security, application system security, middleware security and information service security.

1. **Security problems at the sensing terminal of the Internet of Things**

In the application of wireless positioning in prisons, there exists the problem of leaking the wireless signals and electromagnetic signals of the terminal equipment, such as the electrionic wristbands worn by the criminals, and police tagging, etc. These signals may be intercepted by proprietary equipment and restored to original data, which leads to information leakage; the attacker could also interfere with the wireless signals, which leads to the loss of the information of the wireless positioning system; malicious attackers may even steal relevant information by setting up backdoors or malicious codes on the Internet of Things nodes. At present, in the practical application of prison wireless positioning, transmission data is mainly relative position coordination information, but the impact on security is not yet obvious.

The security measures for the sensing nodes of the Internet of Things in prisons are:

(1) Implementation of electromagnetic information leak protection.
(2) Establishment of password-based identity authentication and access control.
(3) Terminal devices such as electronic wristbands are strictly supervised at the source of production to prevent backdoors at the source of manufacture.

2. **The security problems of sensing networks for the Internet of Things in prisons**

The security protection of the perception network is mainly to solve the authenticity and reliability of the node information of the access terminal, the confidentiality of the user's sensitive information, and the integrity and availability of the data.

The security measures for the sensing network of the Internet of Things in prisons:

(1) Implementation of physical security and confidentiality measures.
(2) The reader collision problems of RFID can be effectively solved through anti-collision technology.

3. **The security problems of transmission network for the Internet of Things in prisons**

The security problems of the transmission network mainly come from the architecture itself and the access methods in prisons, the security problems of information transmission in networks, network attacks across heterogeneous networks, etc.

The main security measures for the transmission network of the Internet of Things in prisons are as follows:

(1) Implement safe access to the terminal of heterogeneous networks.
(2) Ensure that the perceiving data is transmitted securely in the hosted network.
(3) Establish a unified protocol stack and related technical standards.

4. **The security problems of the application system in the Internet of Things in prisons**

These security problems mainly involve specific application business. The main problems are user privacy information and protection, tracking for leaks of information, security of the application system on the Internet of Things, security issues on the Internet of Things, information forensics and protection, hacker attacks, etc.

The security measures for the application systems of the Internet of Things in prisons mainly include the following:

(1) Establish a privacy protection and security mechanism.
(2) Implement data access control and audit trails.
(3) Establish a prison cloud security mechanism.

5. **The security problems of the middleware of the Internet of Things in prisons**

The middleware needs to complete the processing of massive sensing data, and the main security problems are spam and malicious command attacks, the problem of timely processing of data, internal attacks, etc.

The main security measures for the middleware of the Internet of Things in prisons are:

(1) Establish a security protocol of the equipment layer of the RFID middleware.
(2) Implement the access control strategy of RFID middleware.

6. **The security problems of information services of the Internet of Things in prisons**

The Internet of Things platform in prison needs unified security management, which could integrate the above scattered technology and security measures to create a comprehensive effect, as shown in Fig. 3.6.

### *3.3.5 The Applications of the Internet of Things in Prisons*

The system of the Internet of Things in prison is a process of intelligent collection, transmission, unified convergence, and global application centering on prison target information. The starting point of this process is the prison target and the sensing data it produces. The objects of prison supervision are criminals. For the construction of the front-end sensing layer of smart prisons, obtaining criminal data, such as position information, behavior state, and life vital signs, becomes a core task.

The Internet of Things can refine the objects of supervision to an individual, that is, convert an individual into an information node to obtain the behavior characteristics of that information node, such as daily movement tracks, consumption records, and life rules. The data obtained is analyzed through scientific methods, such as differentiation and graphics, and is then modeled to achieve tracking and intelligent judging and researching for the objects in the prison. In fact, current tracking of objects in a

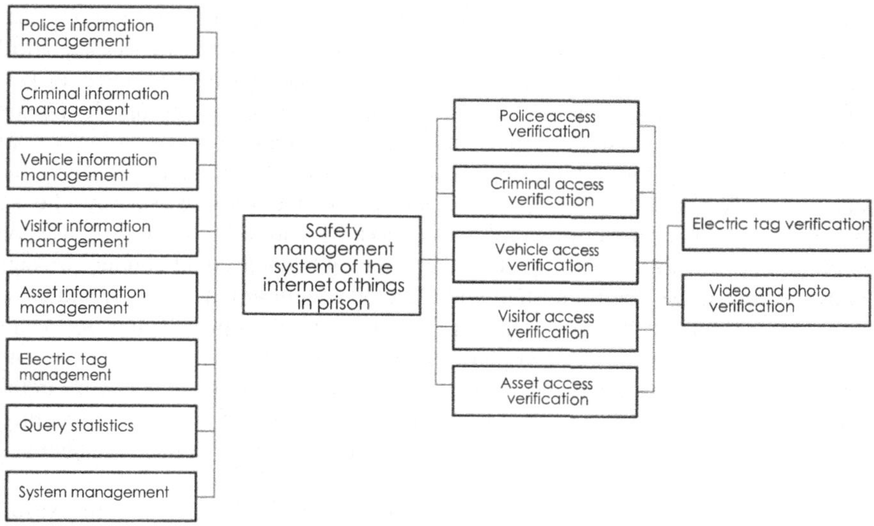


**Fig. 3.6** Functional diagram of the security management platform of the Internet of Things in prisons

prison (that is, wireless positioning in the prison) is achieved by criminals wearing electronic wristbands, which could convert the criminals into information nodes.

Prison video and wireless positioning applications are representative and promising applications of the Internet of Things in prisons. Prison wireless positioning is a technology to achieve the positioning, tracking, and monitoring of specific target positions using satellite positioning systems, RFID radio frequency identification, wireless communication, etc. Through the wireless positioning base station to send the collected electronic tag information to the location server and then through the location server to the computed location data to prison GIS, prison police could intuitively obtain the target location in various forms, such as video display, GIS electronic map, table or report. The main application forms of wireless positioning technology in the prison include the positioning and counting of the persons in the prison (police, criminal, visitor) and the positioning and counting of important working tools and dangerous goods.

With the help of wireless positioning through authorizing or classifying the management levels of the electronic wristbands (active RFID) worn by personnel (such as criminals, polices, visitors, etc.), the prison could limit the access of various personnel in various areas of prisons. Once someone enters the forbidden area, the command center could track he/she continuously, while the video camera monitoring linkage is enabled, which captures images by the monitoring device in the area where he/she is and be shown in the field of the vision of the person who is monitoring. Through observing his/her activities, the person could be judged as to whether he/she has illegal intentions, at the same time, through the coordination between the system and warning equipment, display equipment, access control perimeter, emergency broadcast, and elevator system, once he/she comes to or arrives at the forbidden area, the alarm would sound automatically and prompt the person to leave, or warn the person for being in a restricted area or location in the way of electronic display, or through the emergency broadcast system. At first warning, the person could be notified by voice, or the system could give warning to police who are on duty on this floor, to require him/her to go to the location of the incident as soon as possible. If the above measures fail, the command center could conduct linkage control of the access control system and elevator system at that location and close access to the incident location. The application of wireless positioning in prisons could also improve the security of patrolling police and criminals (such as crossing behavior and attacking behavior), simplify the complexity of the police work in the management of counting the persons, and have the ability of emergency warning, improving management methods of police work, and achieving refined management of distributing special goods or working tools.

With the help of wireless positioning, when finding prisoners who are acting suspiciously, it can be judged by replaying the historical trajectory of the positioning. Even if no abnormalities are found, the analysis of intelligent positioning data could find violations, which could prompt the police to deal with them. Traditional investigation and evidence collection rely on the video monitoring system, and finding the track of a criminal through relaying the video recording could cost much time and effort. If the logic of the track of the person linked with the monitoring video is set, then the corresponding video information could be reorganized and used for evidence collection according to the track route of the positioning.

Compared with the mature outdoor wireless positioning technology (satellite positioning, mobile phone base station positioning, etc.), as one of the current internationally popular technologies, indoor wireless positioning technology needs to be improved for practice in prisons. Currently, it could be a problem to overinterpret the precise positioning in pilot applications in prisons. With the gradual improvement of wireless positioning technology, there is a new possibility to remade the security system of prisons with wireless positioning and video monitoring as double centers. Accordingly, the system of the prison perimeter and access control, etc. should be adjusted, designed, and built with wireless positioning and video monitoring at the core. This could effectively solve the concern of the prison for the 4Ws (who, where, when, what).

The security architecture of the prison with wireless positioning and video monitoring as the double centers is shown in Fig. 3.7.

The nature of the implementation for information construction in prisons is to make the supervision and correction comprehensively and thoroughly normalize and standardize to achieve innovation in prison supervision, to greatly boost the legal and social benefits of prison work, to promote fair law enforcement, and to reduce implementation costs. It is the basic task to reform criminals into law-abiding citizens and to prevent and reduce recidivism. As the central place in the prison, education and correction must be strengthened. Regarding the concept of "People First", the focus

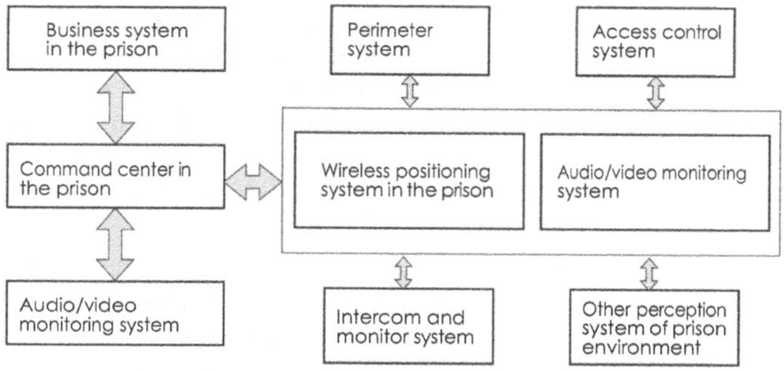


**Fig. 3.7** The security architecture of the prison with wireless positioning and video monitoring as double centers

of the informatization construction must be changed from the traditional "security first" (security informatization) to "correction first" (correction informatization). The current investment in prison informationization is mainly in the field of "security first" but less in the field of "correction first". At present, exploration in the field of evidence-based correction knowledge is an example.

In the construction of the Internet of Things in prisons, the focus of informationization construction must be from traditional "security first" to "correction first". The RFID electronic wristband can be used to tag criminals. The wristband chip could store the basic information of the criminals (including name, age, characteristics, education, occupation before arrest, etc.), crimes (including the nature of the crime, kinds of punishment of the original sentence, sentence, release date, criminal case, etc.), family and social relations (including hometown, relatives, main social relations, etc.), reform performance (including the attitudes of confession, behavior, rewards and punishment, etc.), the corresponding individualized correction module and correction complementation, etc. This could be easily read through mobile office terminasl, such as police mobile Pad, which could effectively improve the ability and effect of the prison education correction.

According to related documents, the RFID electronic tagging system, which was used in the Corcoran State Prison in California in 1997, is an older prison wireless positioning application in America. Its purpose is to track and protect the safety of the personnel working in the prison. The RFID project used in the juvenile offender's correction department in Michigan in 2002, is a known RFID application for wireless positioning of criminals. According to the report, during the 3-year test of RFID wireless positioning of inmates in the juvenile offender's correction department in Michigan, compared with before implementation of RFID application, the rate of escape and violence was reduced by 65%.

The main function of the prison wireless positioning system is to achieve a positioning precision of 0.5–5 m indoors, no error difference between specific rooms indoors and specific areas, no error judgment on a floor, no error difference indoors and outdoors, and a positioning precision of 5–10 m outdoors. The position targets could be divided into personnel positioning and important asset and working tool positioning (including dangerous goods management). The extended functions of the system mainly include automatic roll call, alarm management, video linkage, track playback, and query statistics. Among these, the automatic roll call includes two submodules (counting of persons and important assets and working tools) and two submodules to achieve roll calls based on positioning results and provide an early warning reminder of those who are absent. All the extended functions of the system are based on the positioning database and are realized by reference to positioning results. With the development of wireless positioning technology, positioning precision needs to be improved continuously. As an effective tool of management, in the aspects of counting the persons in time, identifying current locations of criminals, etc., wireless positioning technology provides a great deal of convenience, especially after linkage with wireless positioning and video. When investigating and checking videos, it is quite effective for saving time for police and improving work efficiency. With the development of this sensing technology, motion sensors could be integrated into electronic wristbands to meet the requirements of the prison, such as attitude and acceleration sensors. When there happens to be some changes for the criminals, which is beyond the preset value of the sensor (such as fighting), the data is collected in real time and sent to provide advanced notice to the police. Additionally, the temperature, pulse (blood pressure sensor) could be integrated; when psychological parameters, such as the criminal's temperature and heart rate, etc., are abnormal, the device would send the warning message to management personnel automatically to respond to emergencies (such as violence or sudden death, etc.) quickly.

Currently, the main technology challenges faced by prison wireless positioning applications are as follows:

(1) The problem of energy consumption. With the development of wireless positioning technology, an increasing number of systems will integrate the sensing information of positioning personnel, such as the moving speed, position, height, vital signs (temperature, pulse), etc. However, energy consumption management in a multitask environment is a major challenge. How to reduce the communication cost of node positioning deserves consideration.
(2) The problem of positioning accuracy. Since there may be intersections between positioning base station signals, assuming that base stations A and B are adjacent, when the person is at the overlapping part of the signals, it could be hard to judge which base station the person is in.
(3) The problem of locating many people. Since the processing performance and anti-collision computing, generally speaking, for one positioning base station, the quantity of the messages that the wristbands receive at the same time is limited. If the quantity of the wristbands exceeds what can be processed in the range of the positioning base station, some wristband data may not be received by the positioning base station, which could result in losing signals sent by the radio frequency tag, together with the delay of the positioning response.
(4) The problem of system linkage. To fully function, the wireless positioning system must be deeply integrated and share linkages with other security

systems in the prison. However, the current standards of software and hardware construction in prisons have not yet been unified. Every hardware manufacturer has its own interface protocols, which makes it difficult to integrate various heterogeneous devices in the application of wireless positioning.

(5) The problem of potential false alarms of electronic wristbands. The wristband tag that the criminals wear may be close to some specific objects, such as the wall and floor, causing poor coverage of radio frequency signals. The wristband may not be read, which could result in the absence of criminals in the positioning supervision system and a subsequent false alarm. Including certain specific actions (when the wristband integrates the state sensor), for example, when the person constantly deals while playing cards, could make the system misjudge it to be a fighting incident and give a false alarm.

(6) The problem of positioning of moving nodes. At present, most positioning computing is limited to static networks, so the positioning problem of the moving nodes needs to be further studied.

### 3.3.6 Wireless Positioning Technology Applicable for Prisons

There are many kinds of wireless signals that can be used for positioning, and there are many kinds of wireless positioning systems, including GPS via satellite signals, optical positioning of infrared and laser, sound positioning of ultrasound and sonar, visual positioning of image processing and computer vision, and relative positioning via the principle of gyro.

The international study and application of wireless positioning technology began with the AVL (automatic vehicle locator) system for automatic vehicle positioning in the 1960s. Early wireless positioning technology often used radio waves to transmit signals and detected the signal field intensity of the radio wave transmitted signals between the moving objects and base station, the arrival time and the time difference, the incident angle of the signal and other specific parameters. Then the geometric position of moving objects is determined by related positioning computing. Since the long-wave signal used is easily affected by the weather and buildings, and early wireless positioning mainly focuses on outdoor positioning. Among them, GPS is currently the most successful outdoor positioning technology, but for indoor short-range positioning, the GPS signal cannot penetrate buildings. In view of the special enclosed environment of the prison, in addition to escorting criminals, such as persons going out for medical treatment, arraignments, transfer, etc., outdoor positioning technology could be used, and on most occasions, indoor positioning is applicable. Therefore, the prison wireless positioning technology we discuss in this book will focus on indoor positioning applications.

In the last few decades, indoor local area positioning has had a great market, which has promoted the rapid development of related wireless positioning system research. Infrared positioning, ultrasonic positioning, electromagnetic positioning, and other indoor positioning technologies are emerging.

Infrared waves are electromagnetic waves with wavelengths between radio waves and visible light. Typical infrared positioning attaches an electronic marking on the object to be measured. The infrared IR marking emits modulated infrared rays, which is received by the optical sensor installed indoors for positioning. Although infrared rays have relatively high precision for indoor positioning, infrared rays can only travel in line of sight because light cannot penetrate obstacles. There are two main shortcomings: the line of sight is straight, and the transmission distance is short. This makes the indoor positioning effect very poor, and the average positioning error could reach more than 5 m. When the marking is in the pocket or there is a wall or other coverings, it cannot work well. It is then necessary to install a receiving antenna in every room and corridor, which will be too costly. Therefore, infrared rays are only suitable for short-range transmission, and it is easy to be integrated with fluorescent lights or lights in the room. So this has limitations in positioning.

Ultrasonic positioning adopts the principle of reflection ranging. The system is composed of range finder and some electronic tags that are placed in fixed positions.

First, the upper device sends signals of the same frequency to each electronic tag after being received by the electronic tags. They are reflected and transmitted to the range finder to determine the distance from each electronic tag according to an algorithm. When three or more transponders that are not in a line respond at the same time, the position of the measured object in the two-dimensional coordination system is determined according to the related formula. The overall positioning accuracy of ultrasonic positioning is relatively high, reaching 2–10 cm, the structure is simple, and the power consumption is low. The speed of ultrasonic waves is influenced by air density, and while air density has a close relationship with temperature, in practice, when disturbed by a wall or other obstacles or transmitted in nonline-of-sight, attenuation is obvious, effectiving range of the positioning.

Electromagnetic wave positioning makes use of the electromagnetic waves of wireless propagation in the air. It is currently the most widely used positioning technology of indoor positioning applications. There are many wireless network positioning technologies based on active RFID, WiFi, ZigBee, UWB, Bluetooth, etc. Except for differences in communication mechanisms and frequency bands, these technologies have similar positioning principles, including distance measurement based on TOA (time of arrival), distance measurement based on TDOA (time difference of arrival), distance measurement based on AOA (angle of arrival), similarity fitting based on RSSI (received signal strength indicator), and detection based on proximity relations.

At present, the wireless network technology that can be used in prisons mainly includes the following:

- Radio Frequency Identification (RFID) technology;
- Wireless Fidelity (WiFi) technology;
- Ultrawideband (UWB) technology.

Radio frequency identification (RFID) technology includes active radio frequency technology and passive radio frequency technology.

Active radio frequency is divided into active RFID technology and ZigBee/802.15.4 technology.

Passive radio frequency technology is divided into short-distance RFID technology and long-distance RFID technology (Table 3.1).

1. **Active RFID technology**

- Positioning accuracy: 2–5 m.
- Communication distance: <100 m.
- Radiation intensity: extremely low.
- Three-dimensional positioning capability: None.
- Information security: high.
- Single-point fault tolerance: low.

**Table 3.1** The table of the difference between active RFID and passive RFID

| | Active RFID | Passive RFID |
|---|---|---|
| Tag energy | Built in | From the reader |
| Requirement for the signal strength of the reader | Low | High |
| Communication distance | 0–100 m (adjustable) | Below 3 m |
| Whether it needs to operate manually during the system running period | Every 2 s (or 10 s) send the message, the reader receives automatically and update the system. Automatic system, without personnel intervention | More advanced than bar code, but still needs personnel to close-range scan, to update the system information |
| Information timeliness | Real time, below or equal to 10 s | Wait for the trigger of the reader, the information is as accurate as the last scanning date of the tag |
| Whether it can prevent destruction | Automatic detection of damage alarm | Can't alarm actively |
| Whether it has the function of asking for help initiatively (alarm) | Can integrate the device of call the police for help | Can't alarm actively |
| Boundary control function | Trigger the warning line to auto alarm identification | If across the border, can not be identified quickly |
| Reading accuracy and anti-collision performance | Penetrate the human body, and identify many persons at the same time | No penetrability, when many persons swipe the card, there will appear serious conflict |
| Other functions | Environment monitor (sensor tag)<br>Positioning, tracking of the personnel and the equipment in real-time<br>Access monitoring of the personnel and the equipment | Access monitoring of the personnel and the equipment |

- Positioning principle: similarity fitting based on the signal strength characteristic.
- Features: Active RFID technology positioning requires precollection of the RSSI value of the transmitting signal of various positioning base stations in the area. The positioning device achieves similarity fitting and location determination by collecting and comparing the RSSI values in real time. Because the workload of collecting RSSI data in advance is quite large and the positioning area may change with the movement of the personnel, interference of medium and large electronic equipment can be a problem. This could influence the positioning accuracy. The main working frequencies of these products are ultrahigh frequencies of 433 MHz, microwaves of 2.45.

2. **Short-distance passive RFID technology**

Short-distance passive RFID is the most mature and widely used product in the market. Applications include bus cards, canteen meal cards, bank cards, hotel access cards, second-generation ID cards, etc., and the positioning principle is based on proximity relation detection where the coverage is small. The essence of short-distance passive RFID technology is traditional swiping card technology. Generally, it can be installed beside the key passage where criminals need to swipe the wristband when passing. Local area positioning is achieved with the room as the basic unit through the access control system. The main working frequencies of that kind of product are low-frequency 125 kHz, high-frequency 13.56 MHz, ultrahigh frequency 433 MHz, and ultrahigh frequency 915 MHz.

3. **Long-distance passive RFID technology**

The positioning principle of long-distance passive RFID technology is similar to that of active RFID, but one defect of this technology is that there is no battery supply for passive cards. It requires installation of an electromagnetic base station with larger radiation, whose working energy is supplied by the antenna of the reader, which works continuously. Its single-point capacity is low. Considering the factor of electromagnetic radiation, long-distance passive RFID technology is mainly used in logistics positioning, but it is difficult to apply to person positioning. At present, there are two kinds of long-distance passive RFID technology, working in the frequency of UHF and 2.45 GHz.

4. **WiFi technology**

- Positioning accuracy: 2–5 m.
- Communication distance: <100 m.
- Radiation intensity: low.
- Three-dimensional positioning capability: Yes.
- Information security: low.
- Single-point fault tolerance: general.
- Positioning principle: the distance measurement is based on TOA, distance measurement is based on TDOA, distance measurement is based on AOA, and similarity fitting positioning is based on signal strength characteristics.

- Features: The advantage of WiFi lies in the increasingly popular WLAN. An increasing number of WiFi hotspots will provide more reference points, which will effectively increase the positioning accuracy. For example, Wifi SLAM, an indoor geographic location service provider that has been acquired by Apple, analyzes the signal strength and the unique ID identification code of all the WiFi networks around and uses the reference data set of that area to match, downloading from networks that are already stored in the equipment, and synchronizing the movement of the foot steps through the gravity sensor to achieve the positioning of the user's indoor location. As the current mature technology of wireless IP transmission, because it adopts public wireless network protocols and considering that the well-known encryption algorithms of the wireless network are fragile, WiFi wireless positioning technology has defects in security. However, it may be safer to adopt public protocols, such as WPA2 and WAPI. Limited by the working mechanism of WiFi, the future of WiFi wireless positioning technology depends on the low power consumption technology of the wristband or the improvement of battery capacity.

5. **IEEE802 15 4/ZigBee technology**

- Positioning accuracy: 2–5 m.
- Communication distance: <75 m.
- Radiation intensity: extremely low.
- Three-dimensional positioning capability: None.
- Information security: general.
- Single-point fault tolerance: general.
- Positioning principle: principles that are related to the distance are TOA, TDOA, RSSI, AOA, etc., and those that are unrelated are centroid algorithm, DV-Hop algorithm, Amorphous algorithm, APIT algorithm, etc.
- Features: 820 15 4/ZigBee technology is mainly used for wireless personal area networks (PANs). The characteristics of the network system in the application are short distance, low power consumption, low cost, etc. ZigBee positioning belongs to the wireless sensor network (WSN) technology category, with advanced technical concepts, but its defect is that the network stability needs to be improved because it is susceptible to the environment. ZigBee was originally designed to be used for one-time positioning of every static node location in the field WSN, but it is difficult to use for large numbers of moving objects in the prison. The positioning accuracy of ZigBee will be greatly interfered with by the human body when moving, so reliability is reduced.

According to the positioning principle of ZigBee technology, positioning depends on neighboring nodes in a limited area nearby. When the node moves and leaves that area, the network needs to reorganize (self-organize) and compute the position, so repeated positioning of moving objects quickly consumes the battery capacity of the wristband.

6. **UWB technology**

- Positioning accuracy: 0.1 ~ 0.3 m.
- Communication distance: Determined by the transmitting power, generally <10 m.
- Radiation intensity: extremely low.
- Three-dimensional positioning capability: Yes.
- Single-point fault tolerance: general.
- Information security: low.
- Positioning principle: distance measurement based on TOA, distance measurement based on TDOA difference, combination of TDOA/AOA, etc.
- Features: UWB technology is an advanced wireless communication technology with a bandwidth of more than 1 GHz and no carrier requirements. The information carrier in the UWB radio is impulse radio (IR). Impulse radio is a kind of radio technology that uses the shock impulse (ultrashort impulse) as the information carrier. The feature of this technology is modulating a very narrow (often <1 ns) impulse signal to obtain a very wide bandwidth to transmit the data. Although it is wireless communication, the communication speed could reach more than hundreds of Mbps. Since there is no need for expensive and large moderate-frequency equipment, the UWB shock radio communication system is small in size and simple in structure, has low visibility and confidentiality, has low power consumption, strong multipath resolution, a high rate of data transmission, a strong penetration ability, accurate positioning, and a strong anti-interference capacity. The power spectral density emitted by the UWB system can be very low, so the short-distance UWB radio communication system could coexist with other narrowband radio communication systems. UWB technology represents revolutionary progress for radio technology and has huge potential for broad application wireless communication, radar tracking, precise positioning, etc.

7. **Other positioning technologies**

(1) Wireless positioning technology based on LF low frequency

LF low-frequency (125 kHz) wireless positioning technology is realized by using positioning technology based on proximity detection because the LF low-frequency signal coverage range is limited (approximately 2–5 m). Since low-frequency signals are less affected by environmental change, such as human movement, there is no multipath effect, and the accuracy of identification positioning is high. Currently, solutions for using an accuracy of 125 kHz + 2–45 G trigger positioning are more common in commercial positioning. The essence of this solution is the low-frequency activation trigger technology, which is generally used for semi RFID products (that is, integrating 125 kHz + 2–45 G), as it combines the advantages of active RFID products and passive RFID products. Through transmitting the trigger signals from the 125 kHz positioning base station, once the wristband tag, which integrates 125 kHz, receives the 125 kHz trigger signals, it will wake up (it will be asleep if there are no trigger signals) and play the role of positioning, then 2–45 G could identify and upload the data remotely.

(2) CSS positioning technology

In the field of positioning, IEEE802 15 4a defines two physical layers that could achieve high-accuracy positioning: impulse radio ultrawide bandwidth (UWB) and chirp spread spectrum (CSS). CSS communication is a kind of carrier communication technology, and a Chirp signal is also called a linear frequency modulated signal, which was first applied in the field of radar. Chirp spread spectrum signals have the characteristics of high time resolution, strong anti-multipath ability, high transmission rate, low power consumption, and low system complexity and have good application prospects for indoor target positioning. Unlike the usual sine signal carrier, CSS technology uses a pulse carrier. CSS pulse signals are different from UWB shock pulse signals. UWB shock pulse signals can directly carry information, while CSS technology uses a series of pulses to carry information, which is modulated at the transmitting end, sent out, and extracts the information at the receiving end after filtering and compression.

Compared to UWBs, CSSs have a relatively low frequency and narrow bandwidth, and they are not as good as UWBs with a wide frequency band in terms of positioning resolution and power consumption. The CSS system can be regarded as a compromise version of the UWB system. The positioning effect of the CSS system is slightly worse than that of UWB, but it does improve the cost performance.

(3) Bluetooth positioning technology

Bluetooth technology works by measuring signal strength. It is a short-distance, low-power consumption wireless transmission technology. Wireless access based on Bluetooth technology is referred to as BLUEPAC (Bluetooth Public Access). The topology of the Bluetooth network system has two forms: Piconet and Scatternet. A Bluetooth network consists of a master device and one or more slave devices, which are synchronized with the time and frequency hopping mode of the master device (based on the clock of the master device and the address of the Bluetooth device). Each independent synchronous Bluetooth network is called a piconet. In a piconet, all devices are of the same level and have the same authority.

Bluetooth 4.0 is the latest Bluetooth version in 2012, which solves the problems of power consumption and startup speed, which were widely criticized in the previous version. Bluetooth 4.0-based positioning is the current popular application. For example, iBeacons is a precision micropositioning technology based on Bluetooth 4.0 launched by Apple. When a handheld device is close to a Beacon base station, the device can sense the Beacon signal, ranging from a few milimeters to 50 m. For ordinary Bluetooth (before Bluetooth 4.0), the general transmission distance is between 1 and 10 m, while the iBeacons signals could be accurate to the millimeter, and could transmit to 50 m at the most.

(4) Geomagnetic positioning technology

Geomagnetic positioning does not belong to electromagnetic wave wireless positioning. However, in principle, geomagnetic positioning is close to similarity fitting positioning based on the received signal strength indicator (RSSI). Geomagnetic positioning is based on the objective fact that the strength of the geomagnetic signals is different in various locations of different areas on earth. After recording the characteristics of the geomagnetic signal strength in various locations in the positioning area, geomagnetic positioning is realized according to the similarity fitting algorithm. For example, the InDooRatlas geomagnetic field positioning navigation map system (https://www.indooratlas.com/) is a mobile map application that uses geomagnetism for indoor navigation. It could work in areas without wireless signals. It takes advantage of the built-in magnetometer of the mobile phone to detect the abnormal geomagnetic field and helps the user navigate at the level of reliable 1 m precision without the requirement of other infrastructure. According to the working principle of this technology, when facing environmental change, such as metal object movement in the building, positioning precision would be largely decreased.

(5) Inertial navigation technology

Inertial navigation technology does not belong to electromagnetic wave wireless positioning. Inertial navigation uses inertial elements (accelerometers and gyroscopes) to measure the acceleration of the carrier itself. The three-degree-of-free gyroscope measures the three rotational motions of the carrier; three accelerometers measure the acceleration of three translational motions of the carrier. According to the integration of the measured acceleration signal, the computer computes the speed and location data of the carrier to navigate the carrier. The inertial navigation system is small in size, low in cost, and high in accuracy. The equipment that the inertial navigation system consists of is installed inside the carrier. It does not rely on external information, does not radiate energy to the outside and is difficult to be interfered.

It has good concealment and other features, which may make it possible to replace GPS technology. The disadvantage of inertial navigation technology is that the cumulative error will gradually increase and must be corrected regularly.

### 3.3.7 *Energy Consumption Management of Wireless Positioning*

Taking the energy consumption management of wireless positioning technology as an example, whether the battery replacement of wireless nodes (electronic wristbands) is frequent will determine the application prospects of wireless positioning (pilot) in prisons. How to use energy efficiently to maximize the network life cycle of wireless nodes is a major challenge that wireless positioning technology faces. With the increasing demand for complex tasks, the key to realizing energy savings of the whole system is as follows: first, the node hardware must be designed to support

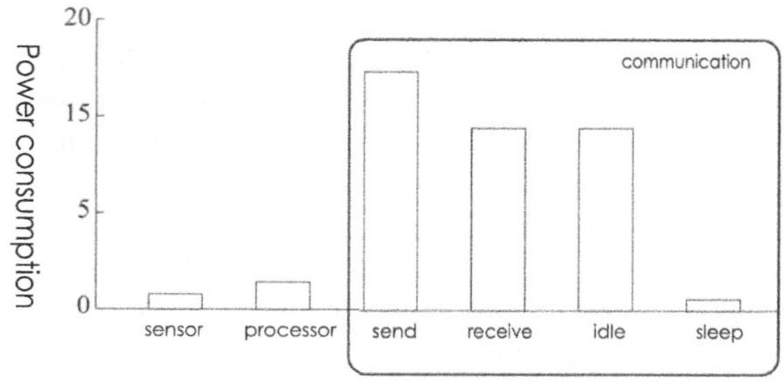


**Fig. 3.8** Schematic diagram of the energy consumption of wireless nodes

the low energy consumption mode; second, the system scheduling strategy must be designed to support the low energy consumption mode. When no task needs to be scheduled, the hardware resources of the corresponding node should be closed to ensure that the whole system is in a sleeping state of low energy consumption. This will reduce or avoid energy loss caused by busy waiting or polling in the common multitask system.

The modules that consume the energy in the wireless node (electronic wristband) hardware resources mainly include sensor modules (integrated state sensing), processor modules and wireless communication modules. The wireless communication module consumes most of the energy, as shown in Fig. 3.8 (see Deborah Estrin's invited report at the Mobicom 2002 conference—Wireless Sensor Networks, Part IV Sensor Network Protocols).

Figure 3.8 shows that when the wireless communication module is in the sending state, the energy consumption is at its maximum. The energy consumption in the idle state and receiving state is similar, which is slightly less than that in the sending state. Therefore, the energy consumption in the sleeping state is minimal.

Using the work mode of the low duty cycle, making the nodes sleep for most of the time could effectively extend the life cycle of the wireless nodes. By setting the time intervals of periodic sampling data, people can reduce the active working time of the processor and wireless communication module. For example, every 100 s could activate 10 s, which could reduce the energy consumption by 90%. Of course, this is at the cost of limiting the actual bandwidth; thus, it objectively shows a delay in positioning speed.

Therefore, the key to realizing energy savings is as follows: when no task needs to be scheduled in the wireless node system, it is important to determine how to make the system enter the sleeping state of low energy consumption and how to avoid energy consumption while waiting and polling.

The realization of the hibernation in a single-task system (only positioning) is relatively simple: task processing in the active state of the processor makes the system enter into a hibernation state in the idle working period of the processor until the wake-up event arrives (the wake-up signal could be set up as needed) to reduce the system consumption.

However, to achieve hibernation in a multitask system, the scheduling mechanism is more complicated than that in a single-task system. For example, in a sensor node that collects and sends data, if only one sensor data point needs to be collected, the sampling period is 60 s per time, and the hibernation time parameter of that thread can be simply set to 60 s. However, in a multitask system, there may be multitasks for collecting and sending data, such as when nodes need to collect temperature sensor data, acceleration sensor data, and carbon dioxide sensor data simultaneously. This may occur if the acceleration sensor requires data collection every 10 s, the temperature sensor requires data every 30 s, and the carbon dioxide sensor requires data every 60 s. This multiform sensing task will make the energy consumption scheduling mechanism of the multitask system more complicated. The system is required to adapt to the interleaving hibernation time parameter in the multitask environment. The sampling frequency and hibernation time that may change dynamically in the application also require the support of the system to respond to changes in the network application environment. For example, in a specific application, the system can dynamically adjust the sampling frequency and hibernation time according to the amount of energy. In addition, some tasks with long-term requirements may wish to delay the sleep of wireless nodes until the end of the task, regardless of other work cycles in the wireless node. The node system is also required to have corresponding flexibility to adapt to this special task behavior. In short, if a mission system wants to satisfy low-power application requirements, it must be clear when the system can safely enter into a hibernation mode.

Energy management of the power supply is a problem that must be considered at all layers of the wireless node protocols. Designing an energy management mechanism on a certain layer may not achieve the overall network energy saving effect. Only by combining the physical layer, MAC layer, routing layer protocol, etc., can one achieve cross-layer optimization as well as comprehensive energy savings.

### *3.3.8 Electromagnetic Radiation of Wireless Positioning*

With the continuous improvement of people's health awareness, the impact of electromagnetic radiation on the human body has gradually become a topic of great concern. Because the effect of electromagnetic radiation on humans is an accumulation process, research also takes a longer period, and people's understanding of it is still very limited. At present, since electromagnetic radiation is intangible, it is easy to make people doubt when no conclusions are available. In addition, some reports on electromagnetic radiation are not objective and scientific and have caused unnecessary misunderstanding and panic.

To determine whether electromagnetic radiation will adversely affect the human body, some aspects should be considered comprehensively, such as the intensity of electromagnetic radiation and the main radiation location and radiation sources. To control the influence of electromagnetic waves on the human body, each country has formulated its own electromagnetic wave protection standards. At present, there are two main standards: one is the ICNIRP (The International Commission for Non-Ionizing Radiation Protection) standard, which is issued by the international commission for nonionizing radiation protection and is mainly applied in Europe, Australia, Singapore, Brazil, Israel, and the Hong Kong Special Administrative Region of China. Another standard is the IEEE standard of the United States, which is mainly applied in the United States, Canada, Japan, South Korea, and Taiwan regions of China. ICNIRP's recommended standards for electromagnetic radiation exposure limits are formulated based on scientific data of thermal effects and immediate effects. The standard puts forward the threshold value for the influence of electromagnetic fields of various frequencies on the human body and then takes 1/10 of the threshold value as the occupational exposure limit and 1/50 as the public exposure limit.

China has set a stricter standard than the ICNIRP exposure limit, with a larger safety redundancy. In 1988, the Ministry of Health issued GB8702-88 "Electromagnetic Radiation Protection Regulations" and GB9175-88 "Health Standards of the environment radio wave in the People's Republic of China". These two standards specified the definition, computation methods, and safe value of the influence of electromagnetic waves on the human body in detail. Therefore, electromagnetic waves are safe for the human body when meeting these two standards. The standard is limited by threshold values of the potential adverse effects of the intensity and frequency of electromagnetic wave radiation on the human body. It separates the standard of environmental electromagnetic waves with permitted radiation intensity into two levels. The first-level standard is a safe zone, which means that all people (including babies, pregnant women, the elder, the weak, the sick and the disabled), who reside, work and live for a long time under the first-level electromagnetic wave intensity of the environment, are in the area that will not be affected by harmful side effects. Newly built, rebuilt or expanded transmitting antennas such as radio stations, television stations, radar stations, etc. must meet the requirements of the first-level standard in public residential areas. The secondary standard is the middle zone, which refers to the area where all people (including babies, pregnant women, the elderly, the weak, the sick and the disabled), who reside, work, and live for a long time under the secondary intensity of the environment electromagnetic, may be exposed to potential adverse reactions. In this area, factories and agencies can be built, but residential houses, schools, hospitals, nursing homes, etc. are not allowed. For structures already built, appropriate protective measures must be taken. The electromagnetic radiation intensity classification standard table is shown in Table 3.2.

Table 3.2 shows that the permitted radiation field of the microwave in the first-level standard safety zone is 10 $\mu W/cm^2$. According to the mobile phone electromagnetic radiation test (http://www.cca.org.cn/jmxf/detail/9034.html) commissioned by the China Institute of Metrology and Measurement and the China Institute of Metrology and Testing. The results show that the electromagnetic radiation of a mobile phone is the largest during the first few seconds when calling and answering. Among the 19 kinds of GSM phones, there are 15 with radiation reaching more than 2000 $\mu W/cm^2$. Some could even reach as high as 10,000 $\mu W/cm^2$, and the radiation can be at least 300 $\mu W/cm^2$, while the wireless interphones are 5 W; When in standby mode, although the mobile phone transmits the regular signal to keep in touch with the base

**Table 3.2** Electromagnetic radiation intensity classification standard table

| Wave length | Unit | Allowable field strength | |
|---|---|---|---|
| | | First level (safe zone) | Second level (middle zone) |
| Long wave, medium wave, short wave | V/m | ≤10 | ≤25 |
| Ultrashort wave | V/m | ≤5 | ≤12 |
| Micro wave | $\mu W/cm^2$ | ≤10 | ≤40 |
| Hybrid wave | V/m | According to the field strength of chief band Distracting wave is weighted by the composite field strength | |

station, the electromagnetic radiation is quite small. Among the 19 tested kinds of GSM mobile phones, the electromagnetic radiation of 11 kinds is below 1 $\mu W/cm^2$ when is in standby mode. Other brands of mobile phones are below 7 $\mu W/cm^2$. At present, electronic wristbands (or electronic footbands) based on RFID, ZigBee, UWB, WiFi and other wireless technologies could meet the first-level standards of the national standards. In this environment, the electromagnetic radiation to a human body is quite small, so it is relatively safe for babies, pregnant women and the older, the weak, the sick and the disabled.

At present, China's national standards related to electromagnetic radiation are as follows:

- GB8702-88 "Protection Regulations for Electromagnetic Radiation";
- GB9715-88 "Hygienic Standard for Environmental Electromagnetic Wave";
- GB12638-90 "Radiation Safety Requirements for Microwave and Ultrashort Wave Communication Equipment";
- GB10436-89 "Hygienic Standard for Microwave Radiation in the Work Environment";
- GB10437-89 "Hygienic Standard for Ultra-High Frequency Radiation in the Work Environment";
- GB16203-96 "Hygienic Standard for Power Frequency Electric Fields in Work Environment";
- HJ/T10 3-1996 "Environmental Impact Assessment Methods and Standards on Electromagnetic Radiation".

# Chapter 4
# Prison Cloud

Cloud computing is the product of technological covergence, including distributed computing, parallel computing, utility computing, network storage technologies, virtualization, load balance, high availability and grid computing. The core concept of cloud computing is to unify, manage, and schedule the networked computing resources to constitute a computing resources pool for service users on demand. The aim of cloud computing is to simplify computing and storage resources for ease of use, such as public water and electricity. Then, the user can operate many kinds of business systems on the cloud platform without the need to build their own dedicated data center. As long as they are connected to the network, the user can easily use and pay for the amount of data traffic used.

Cloud computing associated with a prison can be called the "Prison Cloud", building various business applications on the basis of a cloud platform via virtualization technology and realizing a customized, countable and dynamic allocation and use of virtual servers and storage resources. Applying cloud computing technology could effectively address the problems that currently exist in prison informatization.

(1) Economize construction funds. With the powerful and constantly expanding computing and storage capacity of the Prison Cloud, the capital investment of equipment and management can be reduced. It also saves on upgrading and maintenance costs.
(2) Unify data standard. There is only one standard for all software and data in the Prison Cloud.
(3) Avoid information islands. The data that exist in the Prison Cloud can better realize resource sharing and higher efficiency data transfer.
(4) Assist enhanced security and confidentiality. There is no need to store data locally. The user passes identity authentication and logs in through a unified entrance.
(5) Convenience for unified management and maintenance.



P. Sun, *Smart Prisons*,
https://doi.org/10.1007/978-981-16-9657-2_4

With the gradual enhancement of prison informatization and intelligent construction, each prison will accumulate a large number of supervisions and reform comprehensive business information along with massive prison Internet of Things (IOT) sensing information. It is recommended that the Prison Cloud center be built at the provincial prison management level to collect comprehensive business information such as criminals' penalty correction information at the grassroots level. At the same time, it is also suggested that each prison builds its own cloud center, except for comprehensive business information such as criminal penalty correction that mainly assembles prison IOT sensing information (used to build real big data). This could be summarized at the provincial Prison Cloud platform through prison private networks, data transmission and exchange platforms at any time.

## 4.1 Cloud Computing Architecture

Cloud computing's security and credibility is a hot topic in this industry at present. Establishing a security certificate system of cloud computing requires a process. Generally, prisons are not suitable for public cloud construction, which would allow sharing counting and storage on networks and platforms. A preferable choice would be a private cloud (or customized proper cloud). One can place services with high security requirements on the prison private cloud for management, while the contents of public services can be placed on the public cloud.

The Prison Cloud platform builds many virtual machines on the basis of a physical server, deploying the corresponding application server and database server. This builds a virtualized storage system based on NAS, IP SAN, FC SAN technologies while storing the corresponding application data on the cloud. Using the cloud platform suite to realize a customized, countable and dynamic allocation and use of virtual servers and storage resources builds various prison business applications based on the cloud platform.

(1) Software as a Service (SAAS) is one type of application mode based on the network used to provide users with software service via a browser. It deploys applications uniformly on its own servers, users order application software service on demand from service providers via network, and service providers charge in according to the quantity, time and function of the software user customized, and provides these softwares to users via the browser. From the user's perspective, the expense of the server and software can be saved. From the service provider perspective, the cost of software management and maintenance can be reduced.

(2) Platform as a service (PAAS) is one type of distributed platform that provides a development environment as a service. PaaS is able to provide enterprises or individuals with intermediate research & development platforms, application development, database, application servers, hosting, and application services.

Users can develop their own integrated online application services on the basis of the development environment, server platform, and hardware resources that are provided by service providers.

(3) Infrastructure as a service (IAAS) refers to infrastructure clustered by server as a measured service provided to users. It integrates memories, input and output devices, computing capacity and storage into a virtual resources pool for users while providing the required services. Users can rent the corresponding computing capacity and storage capacity on demand with lower costs, which greatly reduces users' expenditure on hardware.

Synthesizing the above three service modes of cloud computing, the overall architecture diagram of the prison cloud platform, as shown in Fig. 4.1.

A typical example of steps to build a prison cloud platform is as follows:

(1) Build virtual server and virtual machine management platform via VMware vSphere

Sphere is one set of server virtualization solutions launched by VMware. vSphere separates applications and operating systems from bottom hardware, thus simplifying the IT operation. Existing servers and storage can be managed as resources.

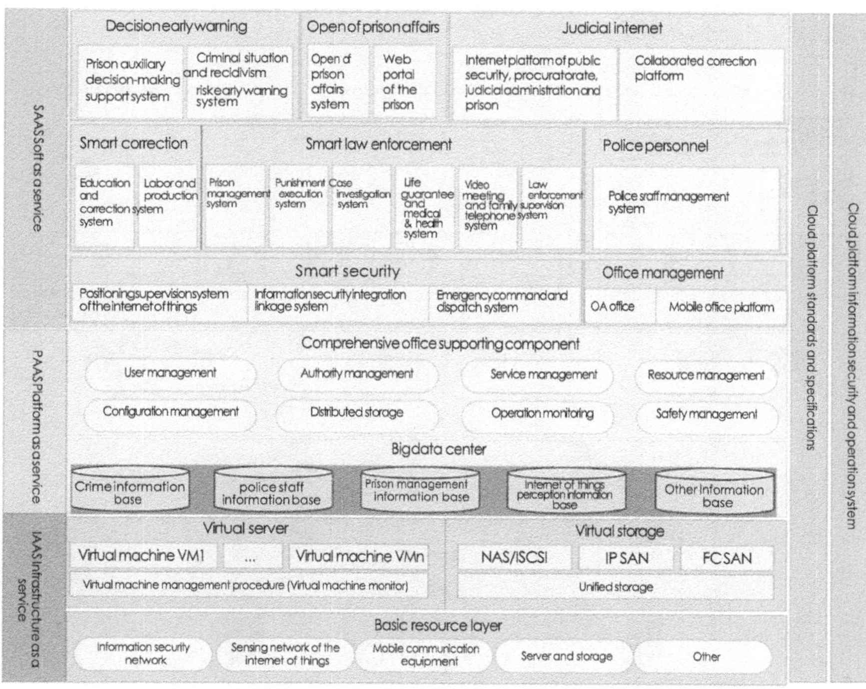


**Fig. 4.1** Overall architecture diagram of the prison cloud platform

(2) Build cloud platform virtualization storage system via network storage technologies (NAS, IP SAN, FC SAN)

Build the unified cloud storage system via network storage technology, and allocate each virtual machine as required to meet the storage needs of application servers and database servers.

(3) Build virtual client via virtual desktop technology

This is remote from a virtual machine in the cloud via a thin client, and applications can be customized for clients on the cloud virtual machine. Most parts of the B/S architecture application can be accessed by a browser and do not need customization. A client displays only the operation interface and the results of the cloud virtual machine, while all applications and storage proceed in the cloud. Taking prison systems into consideration, security must be considered in a special service application; thus, virtual desktop technology is very suitable to adopt.

(4) Build prison business application systems on the virtual server and virtual storage

Realize dynamic management to a virtual machine via virtual management procedures and build prison business applications on this integration for serving users.

The prison cloud platform topology is shown in Fig. 4.2.

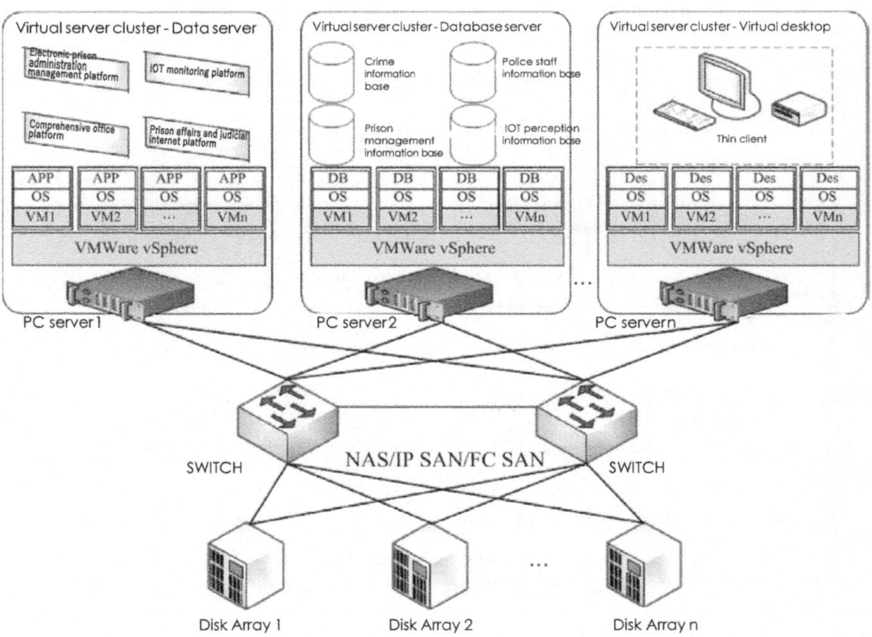


**Fig. 4.2** Topology diagram of the prison cloud platform

## 4.2 Prison Cloud Service

1. **Infrastructure as a service**

Build infrastructure as a service of the prison cloud platform on the basis of IT infrastructure such as prison private network server and storage equipment. This mainly includes a virtual server, virtual storage and infrastructure resources.

(1) Virtual server

Install the virtual machine monitor on the physical server. Build the corresponding virtual server according to the users' actual demand as application server and database server of the deployment business system.

(2) Virtual storage

A clustered cloud storage system was built through mature NAS, IP SAN, and FC SAN technologies. Provide storage for each virtual application server and database server.

(3) Infrastructure resources

Infrastructure resources include the information security network, IOT sensing network, mobile communication devices, servers and storage.

2. **Platform as a Service**

Uniform management to create, load balancing, recycle and backup virtual machines through "cloud platform" management procedures provides corresponding technological components for supporting "cloud platform" applications.

The PAAS of cloud platforms provides technology support for prison cloud platform business applications with commonly used cloud platform management technology components. A big data center contains an inmate information library, police information library, prison management (criminal penalty correction) information library and an IOT sensing information library. Based on the above data, this covers structured, unstructured, and semistructured data with all formats of office documents, texts, pictures, XML, HTML, various reports, images, audio and video information. Due to the difference between structures and sources, different access methods and relevant agreement standards are offered.

(1) User management

For any information system, user management is essential. With a traditional information system, each application system has its own user account and institutional chart. In addition, for the same user, the login account name and password stored are not consistent in different systems, which results in users needing to remember multiple account names and passwords requiring multiple logins when switching between different application systems.

The construction of the Prison Cloud platform should give full play to the cloud's features and advantages and uniform management of the information of users who require logging to the cloud platform application system. This consists of user accounts, login passwords, and prison and prison area information. Cloud platforms take the user account and institutional information in the police personnel management system as standard information, supply it for other application systems by standard web services, and supply synchronization services of user account information. Each application system calls for synchronization service regularly; this ensures that the user account and institutional information of each system are kept consistent with the police staff personnel management system.

(2) Authorization management

Uniform management to all access users' authorization of the Prison Cloud. System users include system administrators, group administrators, general administrators and information inquiry clerks. System users consist of numbers, codes, names, and other elements. In terms of status, system users have two: the locked and the normal. An authorization management module completes the maintenance functions of this system and mainly covers system user additions, system user logout, system user authorization, system user data modification, and system user inquiry; along with role addition, role modification, role authorization and role query; operator group addition, modification, authorization and inquiry; operator addition, modification, authorization and inquiry.

(3) Service management

To comply with the SOA service-oriented architecture specification, the general function of the application system on the prison cloud platform is split into multiple services. This provides service for other application systems by well-defined interfaces and contracts. The benefit is that the entire Prison Cloud platform can be loosely coupled so that each application system can better serve users via continuous evolution.

In essence, the Prison Cloud platform is composed of many services, such as service to deploy virtual machine service, the service to start or close virtual machine, load balance service of virtual machine, information inquiry service, information exchange service, etc. Services management incorporates the following functions:

① Management interface

Provides a complete web management interface and API interface about the service.

② Virtual machine management service

This includes the deployment, start and closure of the virtual machine. Users can fill in virtual application information on a web interface, such as CPU frequency, memory size, hardware volume, the required operating system, database system, application middleware, application software and other information. A cloud platform not only customizes the corresponding virtual machine based on requirements but also gives feedback via its virtual IP address, login account name and password for users.

③ Load balance service of virtual machine

The virtual machine management procedures in the cloud platform can carry out, coordinate and control based on the operating status of each virtual machine. If a physical server runs many virtual machines with large loads and high memory and CPU usage, the virtual machine management procedures will migrate these virtual machines on this physical server to another physical server. If a physical server crash occurs, its virtual machines will be automatically migrated to another physical server.

④ Virtual machine recovery/backup service

To prevent data loss, the system will backup to the virtual machine that runs important application systems, including the application server, virtual machine, and database server virtual machine.

⑤ Information inquiry service

The data of each application system can be provided to platform users or to other application procedures for calling through service.

⑥ Information exchange service

Through the XML data standard, this provides services for data exchange between various application systems on the cloud platform.

⑦ Single Sign on Server

Through deploying a single sign on server, users can login to any application system on the prison cloud platform once, and they will access all mutually trusted application systems.

⑧ Service catalog

Uniform management of web services based on SOA architecture through service catalogs provides a rapid channel for service inquiry and scheduling.

⑨ Log management

The operation log of the virtual server, storage devices, and application system is recorded on cloud storage for viewing and tracking in the future.

(4) Resources management

Resource management is related to the management of physical nodes, such as servers, storage devices, and network devices. It involves the following functions. First is the resource pool; by using a resource abstraction method such as a resource pool, a large number of physical resources can be concentrated into a virtual pool. The second is automatic deployment, which means the entire process automates from creation to use of resources. The third is resource scheduling, which not only makes better use of the system resources but also automatically adjusts cloud resources to assist applications running on it to better cope with sudden traffic, which plays the role of load balancing.

(5) Configuration management

This refers to a template configuration of the virtual machine, key parameter configuration of the cloud platform, and application system function configuration. The configuration management module completes the maintenance of the system database structure and the addition, modification, authorization, inquiry, and database replication functions of system resources. System resources gather all functional modules provided by the Prison Cloud platform application system and functions to incorporate interface menu items, specific modules, module functions, and other contents.

(6) Distributed storage

The distributed storage application of the cloud platform is realized through a distributed storage program.

3. **Software as a Service (SAAS)**

SAAS offers six categories of application services: supervision and security, penalty correction, fair law enforcement, comprehensive support, team building and business collaboration for prison users, based on the architecture of infrastructure as a service (IAAS) and platform as a service (PAAS).

(1) Supervision security

◎ IOT positioning supervision system.
◎ Information security integrated linkage system.
◎ Emergency command and dispatch system.
◎ Criminal Situation and Recidivism Risk Early Warning System.
◎ Prison auxiliary decision-making support system.
◎ Case investigation system.

(2) Penalty correction

◎ Prison Administration Information System.
◎ Penalty execution management system.
◎ Education correction management system.
◎ Labor production management system.

(3) Fair law enforcement

◎ Prison Affairs Open System.
◎ Law enforcement supervision system.
◎ Video meeting and family telephone system.
◎ Prison Portal.
◎ OA office.
◎ Message collaboration platform.

(4) Comprehensive support

◎ Living security and medical and health system.
◎ Prison construction support system.

(5) Team building

◎ Police staff management system.

(6) Business collaboration

◎ The public prosecution department supervises the Internet platform.
◎ Collaborative correction platform.

## 4.3 Judicial Interconnection Based on Cloud Platform

The judicial interconnection system based on the cloud platform is based on current public security, procuratorates, courts, judicial administration, and prison unit informatization. This can accomplish information resource sharing and business collaboration interconnections and strengthen the information resource integration ability between departments. This will meet the support requirements of each unit business operation management platform and promote the construction of the integrated judicial system for handling cases, which provides powerful support for the coordination of the public, procuratorate, and judicial departments.

The overall structure of the judicial interconnection is shown in Fig. 4.3.

Prisons and courts take commutation and parole services to realize collaboration between commutation, parole business, and remote trials. Furthermore, the delivery of criminal cases between the court and the prison is completed. Therefore, it is necessary to build a data transmission system between courts and prisons to (a) realize the information sharing between them, (b) realize the sharing of trial videos and promote the coordination between courts and prisons and (c) realize trial video sharing and promote business coordination between courts and prisons.

Through this information system, prisons and the Ministry of Public Security realize sharing from the population, vehicles, fingerprints, DNA information, information related to the investigation stage, and basic information on criminals, as well as cooperative handling services such as criminal escort, investigation and evidence collection, return to retrial, etc.

The prison and the procuratorate use these information systems to share basic information on criminals, penalty execution information, penalty change information, a criminal's correction information, and prison case investigative information, which further facilitate the procuratorate's supervision of prison law enforcement.

The prison and the judicial administration use information systems to share basic information of parole personnel, information of those released after serving their sentence, and the community corrections information of parole personnel.

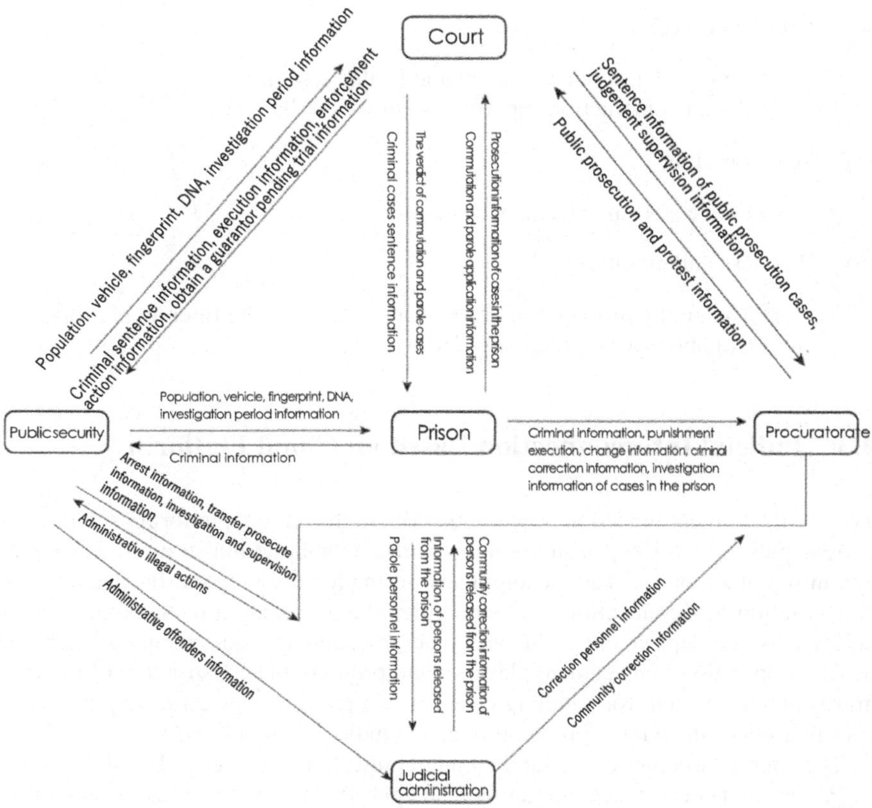


**Fig. 4.3** Overall architecture diagram of judicial interconnection

# Chapter 5
# Prison Big Data

The nature of smart prisons is intelligent processing for prison data. Big data is the smart engine of smart Prisons. Through intelligent information collection, information storage, intelligent data analysis and mining, prison big data application attempts to break the data block segmentation to realize the demand sharing of information resources. This greatly enhances the control level of prison refinement, effectively improves the quality of criminals' correction, and reduces the recidivism rate of those released. The Smart Prison itself is a big data system.

## 5.1 Definition of Big Data

Following the Internet of Things (IOT) and cloud computing, the explosive growth of data has given birth to the emerging concept of big data. Big data have gradually become a technological transformation with far-reaching influence on information and communications technology (ICT). The development and application of big data technology will have a profound effect on the organizational structure of society, the governance model of the country, the decision architecture of the company, commercial business strategies, and individual lifestyle.

The scale of the data involved in big data is so vast that it is almost impossible to capture, manage, breakdown and organize the information to help enterprises make business decisions using current mainstream software tools. The data scale of big data is beyond the scope of the traditional database software collection, storage, management and analysis ability, referring to various data sources, various data categories and formats.

The basic characteristics of big data can be understood from several aspects, including scale, change frequency, categories and value density. Generally, big data characteristics can be described as 5 V, that is, volume, variety, velocity, value and veracity.



P. Sun, *Smart Prisons*,
https://doi.org/10.1007/978-981-16-9657-2_5

◎ Volume: The scale of data aggregated for analysis is enormous. "Big" is the relative concept. There will be large differences in magnitude in different cases. The starting measurement unit for big data is at least PB (1024 TB), EB (1024 PB) or ZB (1024 EB).
◎ Variety: The data form is diverse and can be divided into transaction data, interactive data and sensing data based on the generation type; social media, sensing data and system data based on the data sources; texts, picture, audio, video and geographical location information based on data formats; sturctured, unstructured and semistructured based on data relationships, among which unstructured data is becoming increasingly important in big data. Data owners are divided into enterprise data, government data, social data, etc.
◎ Velocity: With the fast growth rate of data, while there is a requirement for speed and efficiency of data access, processing and delivery.
◎ Value: Although we have massive data and the underlying value behind big data is huge, the value density is low. It is suggested that a powerful machine learning algorithm needs to be used to quickly complete the extraction of values. In addition to the foundation value, "primary value" of the data itself, this can also analyze and find the derived "secondary value" in the data. The focus of big data lies in data analysis and application.
◎ Veracity: On the one hand, massive data under a virtual network environment need to take solutions to ensure its reality and objectivity. This is the urgent need for Big Data technology and business development. On the other hand, using the analysis of big data, reestablishing and predicting the true nature of things and future development trends are also key issues for the development of big data.

## 5.2 Cloud Computing and Big Data

Cloud computing as the name implies emphasizes "computing", i.e., its computing capacity (data processing capacity). The aim is to save IT deployment cost. Big data emphasizes "data", i.e., the object of computing (data storage capacity). It seeks out the value in data. Both are dynamic and static relationships. Big data requires the ability to process Big data, which means a powerful computing capacity. The key technologies of cloud computing, such as mass data storage and management, are the foundation of big data technology. The Internet of Things, cloud computing and big data are shown in Fig. 5.1. Obviously, prisons need to pay attention to the construction of the sensing data collection terminal for the Internet of Things and customize a data layout in advance for intelligent applications such as "Prison Brain" (Figs. 5.2, 5.3 and 5.4).

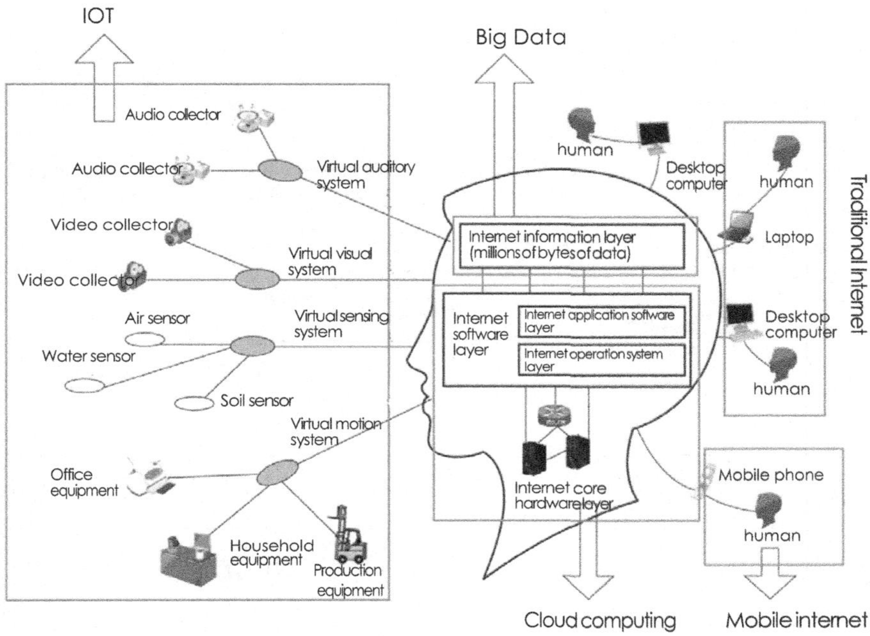


**Fig. 5.1** Internet of Things, cloud computing and big data. *Note* The original image is taken from the book Internet Evolution. Liu Feng, Tsinghua University Press, 2012

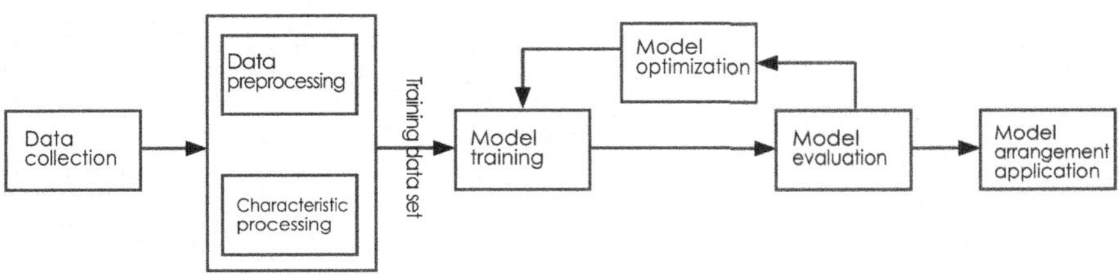


**Fig. 5.2** The general steps for machine learning

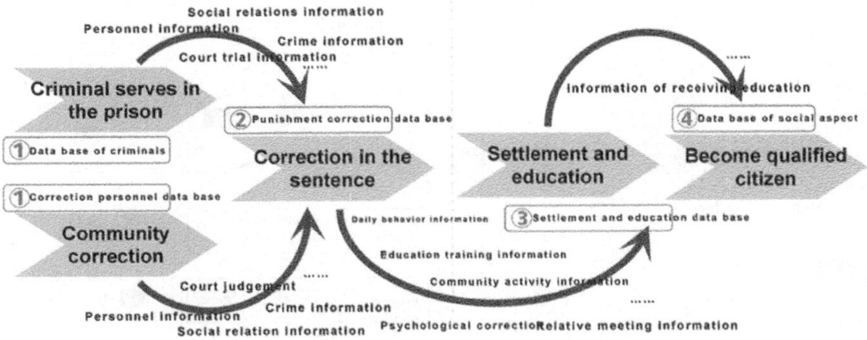


Fig. 5.3 Full cycle data on criminals

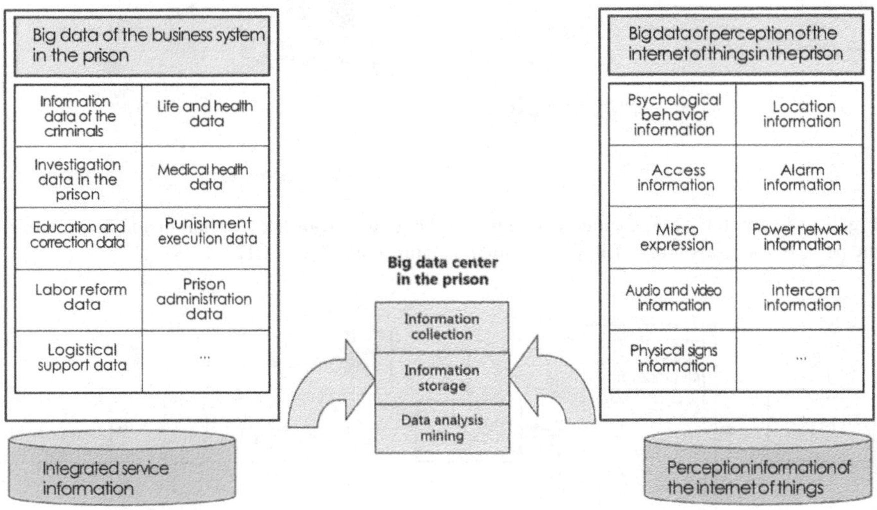


Fig. 5.4 Prison big data sources

## 5.3 Artificial Intelligence Technology and Big Data

Artificial intelligence can be simply understood as "offering human intelligence for machines". This means creating a system that is able to collect data, preprocessing the data, and building model assumptions to predict or influence the results. Since "artificial intelligence" was put forward at the Dartmouth Conference in 1956, it has gone through more than 60 years of ups and downs. This clearly shows the development history from "reasoning" to "knowledge" to "learning" as the focus. At present, artificial intelligence can only fulfill some specific tasks, showing that we are still in the early stages of artificial intelligence.

Machine learning is a crucial way to realize artificial intelligence. It mainly studies how the computer understands human thinking to acquire new knowledge and skills and simulate and realize human behaviors. Humans learn according to experiences, whereas machines discover routines from existing data samples and find the mathematical model of those samples and then use this model to predict unknown data. The commonly used typical machine learning methods include regression analysis, classification (such as decision trees, neural networks, support vector machines), clustering (such as K-means), data dimensionality reduction, feature extraction, etc. Machine learning can be classified into supervised learning, unsupervised learning, semisupervised learning and reinforcement learning according to whether the training data are labeled. Machine learning plays a vital promoting role in the development of machine learning.

The following are the steps to completing a machine learning mission: data acquisition, data preprocessing and characteristic processing, model training, model evaluation and model optimization as well as model deployment and application.

With the accumulation of big data, the improvement of hardware computing capacity and the enhancement of algorithm modeling, AI technology based on deep learning has developed rapidly in the twenty-first century. The most representative is the man–machine battle of AlphaGo in 2016. The top expert of Human-Go was defeated by the deep-learning AlphaGo. Deep learning is the most important branch of machine learning. This imitates the mechanisms of the human brain by building a neural network that simulates human analytical learning. The advantage of deep learning is that feature extraction is performed automatically without manual work, effectively obtaining knowledge of the essential features of data from the sample data. Deep learning is also classified into supervised learning and unsupervised learning; for instance, CNNs is a supervised machine learning model.

Data and knowledge are the two conerstones of artificial intelligence. How do we integrate data intelligence with knowledge intelligence? In 2012, Google released the knowledge search product "Google Knowledge Graph", and the knowledge graph has become the general name for all kinds of structured knowledge systems and is commonly used for intelligent tasks such as information retrieval, reasoning and decision making. The integration of knowledge and deep learning is currently a promising study direction.

Using the concept and technology of big data has been mentioned before, which extracts data from the application systems to achieve the path of "data—information knowledge—intelligent". Therefore, in prison video big data applications, whether high-definition cameras are used or whether the ability to see clearly is good is not neccessarily more important than machines picking up prison video knowledge automatically. As a large amount of video data is presented, video observers will be overwhelmed. This is the opposite of the original design of video surveillance. Currently, the combination of machine learning and big data has contributed to great value, and big data analytical technology has been commonly used in many industries and fields.

## 5.4 Prison Big Data Application

After many years of construction, prison informatization accumulated a certain amount of basic business system data. With contracts for data storage records and annual queries in quantity of the National Crime Information Center (NCIC), it can be predicted that under the conditions of personal privacy and public security, a comprehensive criminal database that collects fingerprints, human faces, irises, DNAs, and voices can be established. The interconnection of information between departments of the Public Prosecutions and Law Department and the application scope and frequency of data will be further enhanced, especially the development of prison IOT applications. The superlarge, heterogeneous and diversified real-time IOT sensing data will prepare for the application of big data in prisons.

From the perspective of big data aggregation, prison big data sources are divided into two categories:

(1) Comprehensive business system information on prison penalty correction with a focus on static data.
(2) Prison IOT sensing information collected by sensors with a focus on dynamic data.

### *5.4.1 Big Data Application for the Comprehensive Business System Information of Prison Penalty Correction*

The comprehensive business system information of penalty correction refers to OA office business system information, penalty portal information, prison administration management business system information, penalty implementation business system information, education correction business system information, prison labor business system information, in-prison case investigation business system information, living allowance and medical treatment and public health system information, police staff personnel management system information, video meeting and family telephone system information, prison affairs disclosure system information and law enforcement supervision system information, etc.

Prison business systems produce a large amount of business data and statistical information every day. The primary value focuses on recording daily business activities and offering evidence for handling commutation and parole. It is a means of prison management. Its secondary value lies in discovering various types of information, such as the relationship between criminals' daily appetite and correction effectiveness from various business data. This helps prisons improve the safety of supervision and the quality of reform to greatly promote the prison governance level.

### 5.4.2 The Big Data Application of Prison IOT Perception Information

The prison IOT sensing information collected by sensors involves various security equipment and facilities, such as video surveillance, AB doors, access control, perimeters, power grids, radio frequency equipment, monitoring intercoms and other security systems. In addition to the boiler, generator, firefighting equipment and other prison equipment. Through the sensors mounted on different security facilities and equipment, a large amount of dynamic sensing data is collected, such as video, opening and closing doors, changes in grid voltage and including real-time location information and vital sign information generated by criminals' RFID electronic wristbands. These data can not only judge whether a boundary is crossed but can also expose the behavioral habits of prisoners, potential escape risks, and other secondary values.

Video surveilliance is a vital component of the prison security system. Each prison's key areas are mounted with surveillance cameras generating a high volume of surveillance videos all the time. These videos can not only be used to monitor the real-time activities of prisoners but also be used to assist in analyzing the prisoner's behavioral trajectory, intention, and other secondary values. How to improve the security assurance of surveilliance locations through intelligent analysis of surveillance video has become one of the important directions for the application of big data technology in prisons.

Take video intelligence analysis technology as an example. Prison application scenarios are particularly concerned with any abnormality in video surveillance to avoid all kinds of security risks. Using the traditional tactics of people staring at surveillance screens, polling in the face of a large number of camera surveillance images proves difficult in finding problems on time. Video surveilliance equipment is often used as a "video recording system" for retroactive or forensics collection. How to change video surveilliance from passive response to active warning, from evidence collection to precaution in advance, has great significance for the application of prison video surveilliance.

Video intelligent analysis is an intelligent surveillance technology based on target behavior. It is the branch of computer image vision technology application in the field of security. Unlike traditional video motion detection (VDM) technology, video intelligent analysis separates the target from the background in the scene, identifies the real target, eliminates background interference such as leaf shaking, water waves, and light changes, and then analyzes and tracks the target behaviors occurring in the camera scene. The natural difference between intelligent video analysis and motion detection is that the former can accurately identify the real active target in the videos, while the latter can only judge the changing content of the video and cannot distinguish the target and background interference. The core of so-called intelligence is target recognition and behavior analysis. Compared to motion detection, intelligent video analysis has improved the anti-interference ability in a qualitative way, especially in detection and recognition in harsh environments such as low illumination, heavy rain, heavy snow, thick fog, strong wind, hail, rain or sunshine, or similar sudden changes.

Video surveilliance data are currently the largest IOT sensing data generator in prisons. The volume of data produced by video is huge. A normal SD camera can produce up to 7 GB of data a day, while HD 720P and 1080P are even larger. The massive video content brings great challenges to data analysis and retrieval.

For prison video surveilliance systems to achieve prewarning and active prevention, this depends on whether the timeliness of the algorithm in the big data environment can meet the requirements of prisons. From the perspective of the current development of big data, technologies suitable for big data, including massive parallel processors (MPPs), distributed file systems, distributed databases, cloud computing platforms, and scalable storage systems, are all based on the ability to efficiently process large amounts of data during the required period.

A common big data application framework for video surveillance is shown in Fig. 5.5.

The core of the video surveilliance big data application framework is a big data platform for data storage and processing. A big data platform is realized based on the Hadoop cluster through the unified data access and storage of the video surveilliance equipment data and other IOT sensing equipment data. Various types of structured and semistructured data will be stored in HDFS and HBase through cleaning and transformation. Hive is used for data extraction, transformation and loading (ETL).

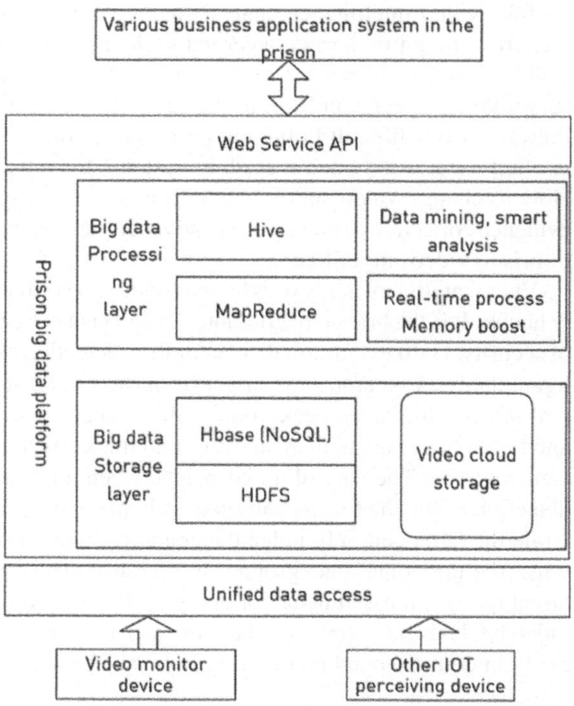


**Fig. 5.5** Application framework of video surveillance big data

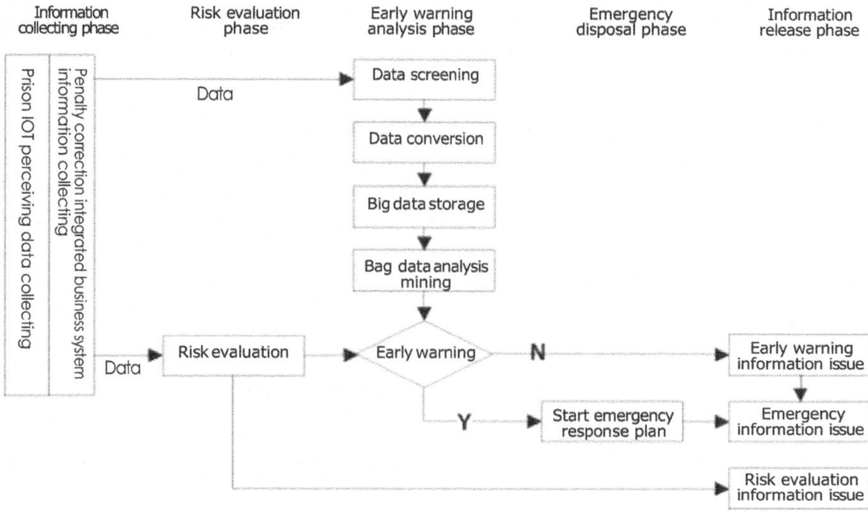


**Fig. 5.6** Procedures of risk early warning decision-making based on big data

MapReduce can conduct large-scale parallel processing and analysis of data stored in HDFS and HBase.

Through big data technology, we can further identify the valuable information behind the massive video surveilliance data and quickly give feedback to the risk early warning decision-making system. This would be the developing direction for future video surveillance applications.

Risk early warning decision-making-based big data can be classified into 5 phases: information collection, risk evaluation, early warning analysis, emergency response, and information release, as shown in Fig. 5.6.

The procedures of risk early warning decision-making based on big data are as follows:

(1) Information collection phase. This collects detection index data and risk assessment index data required for the early warning phase.
(2) Risk evaluation phase. This evaluates the index data of risk assessment.
(3) Early warning analysis phase. This conducts prehandling against detection index data, including data screening and data transformation, filtering out data that does not meet the requirement and storing them in the database. The analysis results are generated by using data mining methods.
(4) Emergency disposal phase. This is to judge whether the early warning threshold is met based on the data analysis and mining results at this phase and the risk evaluation results in the risk evaluation phase. The emergency plan can proceed in case of early warning.

(5) Information release phase. This corresponds to the release of early warning information, emergency information release, and risk assessment information release.

The combination of big data technology and video surveilliance can automatically detect abnormal prisoner behavior in a timely manner through data mining, intelligent analysis, real-time processing, memory acceleration, and video intelligent analysis technology. In addition, it provides an early warning in a timely and reliable manner, thereby effectively making up for inadequate police forces. This contributes to a profound change in the field of prison video surveilliance business.

### 5.4.3 *Prison Comprehensive Big Data Application*

The framework of prison comprehensive big data applications is shown in Fig. 5.7.

1. Data source

The data source refers to data that prisons need to summarize in the data warehouse. It includes business data and IOT data coming from each department of the prison. The detailed contents are shown in Table 5.1.

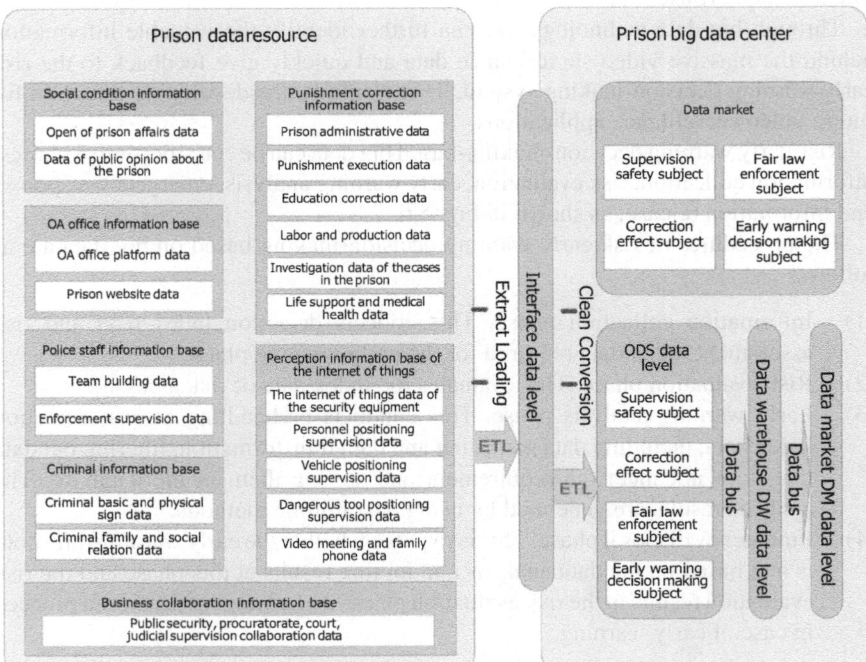


**Fig. 5.7** Prison big data application framework

**Table 5.1** Data source

| Information base | Data set |
|---|---|
| Social information base | Prison affairs public data, prison-related public opinion data |
| OA office information base | OA office platform data, Prison website data |
| Police staff information base | Team building data, law enforcement supervision data |
| Criminal information base | Criminal basic and physical data, criminal family and social relations data |
| Penalty correction information base | Prison administration data, penalty execution data, education correction data, prison labor data, prison case investigation data, life security and medical and health data |
| IOT sensing information base | Security equipment IOT data, personnel positioning supervision data, vehicle positioning supervision data, dangerous tool positioning supervision data, video interviews and family call data |
| Business collaboration information base | Collaborative data of public security, procuratorate, and judicial department |

2. Interface data layer

The interface data layer is a stored data space established to ensure successful processing of data transformation. It is a temporary storage area where prisons share raw data and access the big data center. The interface data layer provides the service of the basic data platform exchanging data with the outside. The existence of the interface data layer provides the most original data benchmark for data verification of the data warehouse. First, the shared library data is quickly copied into the interface data layer. Then, this data layer is washed, transformed and mapped by complicated data processing, and finally, it is transferred into the ODS data layer. The entire process of data processing is traceable and verifiable (Figs. 5.8, 5.9 and 5.10).

The role of the interface data layer is mainly reflected in the following three aspects:

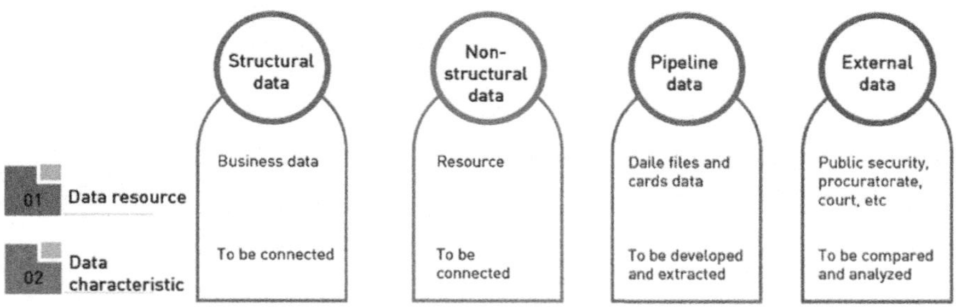


**Fig. 5.8** Smart prison data source and data characteristics

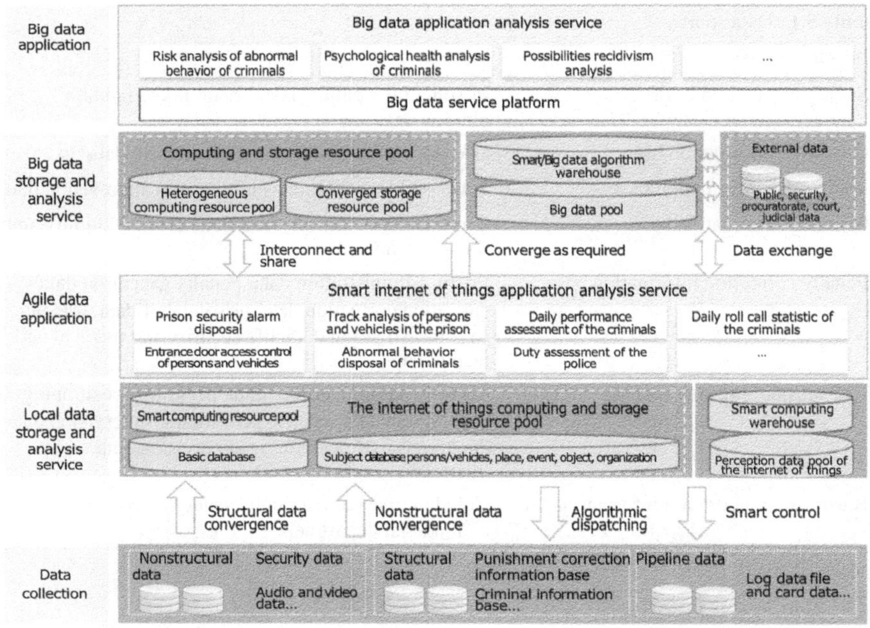


**Fig. 5.9** Smart prison global data construction plan

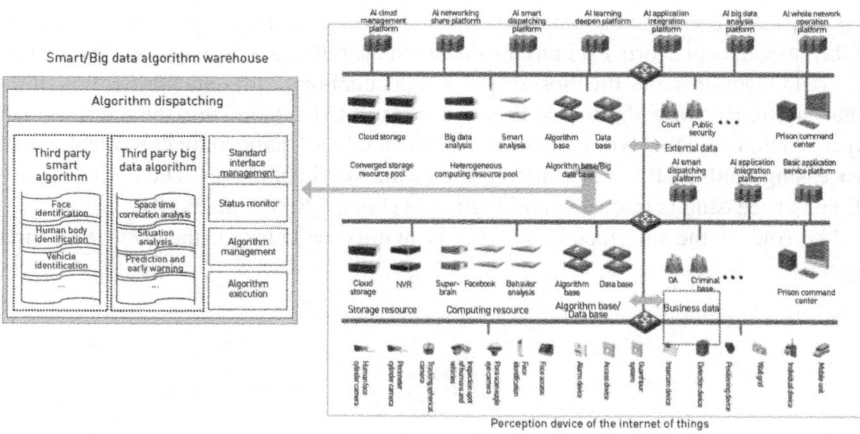


**Fig. 5.10** Smart prison big data algorithm warehouse

◎ Reduce the occupation of resources within the prison business system and avoid affecting prisons business systems by complicated data transformation.
◎ Transform the different forms of prison source data (database, XML, text, etc.) into a unified storage form.
◎ If there is a system failure in the ODS layer, data recovery can be conducted directly from the interface data layer without re-extracting from the shared library.

3. Operational Data Store layer

As a transition from database to data warehouse, the ODS data layer stores and manages the detailed data of prison history and current business that is obtained from the interface data layer through an ETL process (extracting, transforming and loading from source to destination). ODS data come from each operational database of the prison business system. The data structure of ODS is generally consistent with the data source. Additionally, there will be some consolidation after data cleansing and processing. ODS plays a connecting role in the three-layer data architecture of DB&ODS&DW. The interface data layer to the ODS data layer is to complete the ETL processing.

4. Data Warehouse layer

The data warehouse is an integrated, relatively stable, large data storage collection that reflects historical changes. It is designed for analyzing data and used to support management decision-making. The data in a data warehouse is organized based on certain subject areas, and it eliminates the inconsistency of the source data. OLTP (on-line transaction processing) is normally performed in traditional databases. OLAP (on-line analytic processing) is processed in the data warehouse. The data layer of the data warehouse summarizes the integrated comprehensive data. The data of a data warehouse is only allowed to be added but not to be deleted or modified. The data in ODS is "real-time value", whereas the data in a data warehouse is "historical value". The data warehouse mainly provides query services, deletion and modification in distributed systems.

5. Data Mart layer

For the specified application purposes, the data of the DM layer is aggregated from detailed data of the DW layer. It can also be called department data or subject area. The DM layer serves the specified application subject of prisons. It can be flexibly customized based on user needs to meet the requirements of prison statistics, reports, early warnings, performance appraisals and so on. Generally, each prison business application has its own DM. According to the government objectives of prisons, 4 core themes are listed in Fig. 5.11: supervision safety, correction effect, justice of law enforcement, and early warning decision-making.

6. Data mart

The data mart is an integrated, theme-oriented data integration designed to support the function of decision support systems. It is the logical concept of the data center,

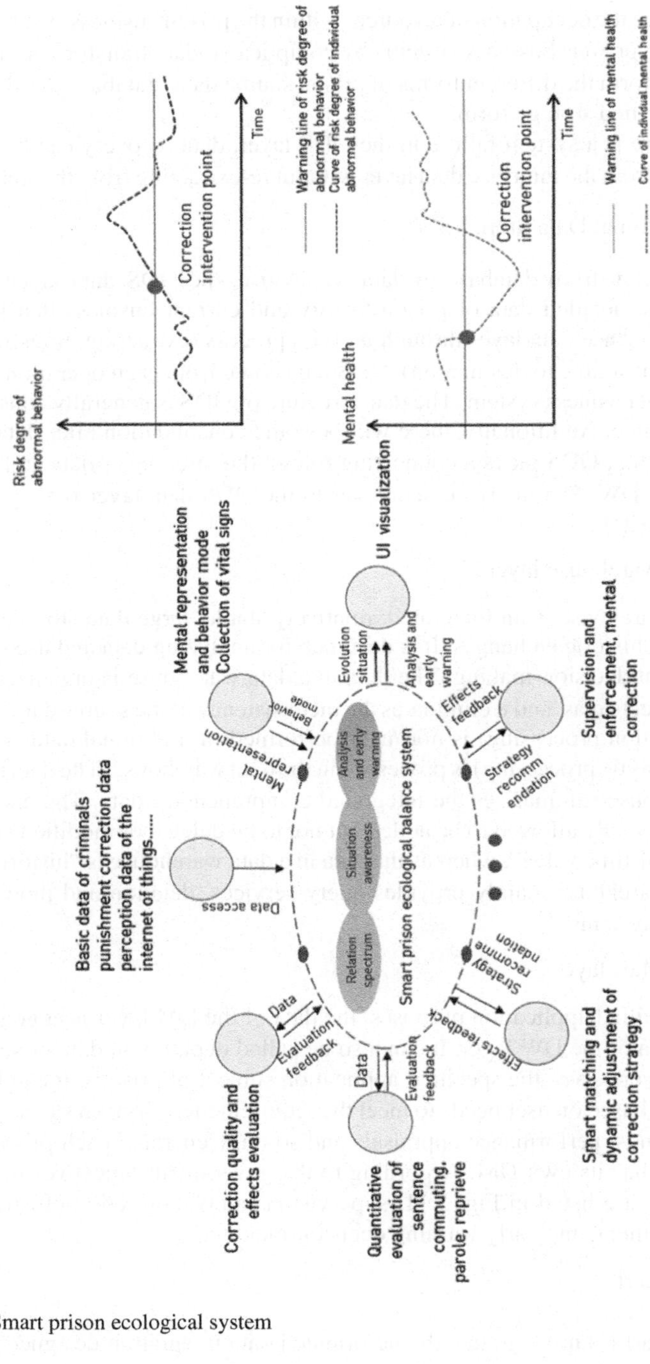


**Fig. 5.11** Smart prison ecological system

an integrated data opened for data applications and the access of data applications. The implementation process of a data warehouse commonly starts from a department data mart and forms a completed data warehouse with a number of data marts.

## 5.5 Smart Prison Construction

Smart prison construction is the iterative upgrade of informatization prisons and pays more attention to data governance, system fusion, and intelligent upgrades. It uses scene-based multibusiness collaborative applications as a starting point to promote prison security risk monitoring of early warning barriers moving forward. It promotes an emergency command system to be intelligent and efficient, robustly builds a "prison brain", and strives to build a "holistic and intelligent governance" of prisons.

(1) Reconstruct the standardized working mechanism of prison digital transformation

This builds a two-wheel driving service system of "standardization + informatization" based on the principles of "standardization leads informatization, informatization promotes standardization". "Standardization" does not mean a standard system of traditional prison management, but it establishes clear rights and responsibilities and a unified coordinated digital transformation standardization work mechanism. This also means building a well-structured and clear prison digital transformation standard system with high system efficiency. During the transition, the double superposition of the standard system of traditional prisons management and the standard system of prison digital transformation are major reasons that make it difficult to realize an informatization dividend.

The informatization contruction route of a prison is traditionally led by prison administration, prison investigation, and other business departments commonly centered on management. The construction of various management information systems in prisons is used to obtain the "data" required by management, which usually adopts ledger reports with "Form-filled" collection. Most data collected by "Form-filled" belongs to the results type, but it is hard to reflect the business process of prisons by just looking at these results, whereas the refinement of management during the process is precisely one of the characteristics of smart prisons. Thus, the data collected by a "filling form" process is prone to distortion, lag, and omission. The logical starting point of smart prison reconstruction must instead be based on the perception of the IOT. That is, it should come from the perspective of service, people-oriented, and automatic data collection based on "perceived" to comprehensively enhance perception.

(2) Smart Prison Comprehensive Data Construction Plan

The data collection side (data collection layer) realizes precise edge perception. The prison side (local data storage and analysis service layer, agile data application layer) realizes efficient on-demand aggregation and an agile edge response. The prison management side (big data storage and analysis service layer, big data application layer) realizes intelligent reliable multidimensional analysis and standard compatibilities.

◎ Data collection layer

This collects the structured data of prisons (such as penalty correction information base, data of criminal information base), unstructured data (such as security data, video data), stream data (such as log data, file card data), external data (such as collaborative data from the Public Prosecutors' Office).

◎ Local data storage and analysis service layer

This converges the structured, unstructured, and stream data of data collection layers into the local IOT computing storage recource pool, including basic databases, thematic databases (human/vehicles, places, events, things, organizations), intelligent computing resource pools, IOT sensing data pool, and an intelligent algorithm warehouse.

◎ Agile data application layer

This provides an intelligent IOT application analysis service. It covers prison security alarm disposition, human and vehicle trajectory analysis in prisons, prisoners' daily performance evaluation, prisoners' daily naming statistics, gate entry and exit control of people and vehicles, prisoners' abnormal behavior disposition, police duty performance assessment, etc.

◎ Big data storage and anlysis service layer

This refers to a big data pool (data is on-demand aggregated by the agile data application layer), an intelligent/big data algorithm warehouse (architecture as shown in the figure below) and offers interconnected and shared computing storage resource pools (heterogeneous computing resource pool, converged storage resource pool) for other layers.

◎ Big data application layer

This provides big data application analysis service based on a big data service platform. It includes risk analysis of abnormal behavior of criminals, mental health analysis of criminals, possibility analysis of reoffending.

(3) Smart prison ecological system

Smart prisons themselves can be seen as continuously developing ecological systems. "The evolution of criminal psychology and behavior patterns" and "Intelligent matching of correction strategies" are a pair of balanced factors in the smart prison ecological system. The key purpose here is to "dynamically adjust to achieve the best correction strategy" to solve the many changes in prison criminals' psychology and behavior, where correction strategies are hard to adjust and follow up at appropriate times. This core issue of regulatory correction focuses on the "corrective intervention point" before the destruction of ecological balance. As shown in the following figures, in the era of prison informazation, issues such as the correction case knowledge database and personalized correction strategies are addressed. The era of smart prisons focuses on solving real-time issues such as intelligent matching of correction strategies and dynamical adjustments.

The natural difference between smart prisons and informationalized prisons is that the former will realize procedural reengineering, rule reconstruction, function shaping and ecological construction in all fields of the prison to build a fair and just law enforcement and supervision system and a comprehensive intelligent collaborative governance system.

# Part III
# Smart Application

# Chapter 6
# Prison IOT Application—Vital Signs Monitoring System

## 6.1 System Overview

Prison vital signs monitoring systems are derived from the prison's critical need against real-time perceptive physiological parameters such as a criminal's temperature, heart rate and blood pressure. Psychological stress can have a variety of impacts on the human body, including heart rate, blood pressure, facial microexpression, sleep state, and temperature changes. The development of IOT radio frequency sensing technology has been able to monitor these physiological parameters from a certain distance. Although reliability is expected to be improved, in terms of practical application, it may indicate that there is urgent violence, suicide, sudden death and other events when criminals have abnormal physiological parameters.

Vital signs are the key indicators used to judge whether a human is behaving normally. This refers to breathing, temperature, pulse and blood pressure, known as the four major indicators in medicine. They are critical to maintaining the body's normal activities, where any abnormalities may lead to serious consequences.

1. **Body Temperature**

Body temperature refers to the temperature inside the human body. The normal armpit temperature is 36~37 °C, which might fluctuate less than 1 °C within 24 h. Physiologically, it is slightly lower in the morning and slightly higher in the afternoon, or after exercise and eating. In addition, body temperature is affected by gender, age and physical activities. It could rise with mental tension and emotional excitement.

Sometimes it can rise approximately 2 °C under certain stressors. The temperature of each part of the body is actually different and can be divided into two parts: the deep temperature and the surface temperature. The terms "deep" and "surface" have no strict anatomical meaning.



P. Sun, *Smart Prisons*,
https://doi.org/10.1007/978-981-16-9657-2_6

1. Deep temperature. The deep body temperature is relatively stable and even. Due to different metabolic levels, the temperature of each internal organ is slightly different. Continuous blood circulation will keep the temperatures of different organs approximately the same. Theoretically, body temperature refers to the blood temperature in deep parts of the body. It can represent the average temperature of the body's internal organs.
2. Body surface temperature. The body surface temperature is lower than the deep temperature. There is an obvious temperature gradient from inside to outside of the body. Body surface temperature has a certain thickness, which plays the role of a heat insulation layer in temperature regulation, through which to maintain the relative stability of the deep body temperature. The outermost layer of the body surface is the surface of the skin. Its temperature is called skin temperature. Skin temperature varies greatly in different parts of the body. When the ambient temperature is 23 °C, the skin temperature of the forehead is 33~34 °C, the torso is 32 °C, the hand is 30 °C, and the foot is 27 °C. Under cold conditions, the skin temperature of the extremities (hands and feet) decreases significantly as the temperature drops, while the skin temperature of the head changes relatively little. The skin contains many vessels, and factors affecting the skin vascular systole can change the skin temperature.

Currently, remote, multitarget and noncontact infrared thermal imaging thermometers used to identify feverish crowds in public places are relatively mature, safe, and convenient.

2. **Heart rate (Pulse)**

Pulse is the rhythmic and periodic undulation of the arterial wall in the cardiac circle. Pulse is normally checked by bilateral radial arteries. The number of normal pulses is consistent with the number of heartbeats, with even rhythms and equal intervals. The blood circulation is accelerated, and the pulse is faster due to daytime activities; the pulse is slower due to less activities at night. Infants can be of 130~150 times/min, children are of 110~120 times/min, normal adults are of 60~100 times/min, the elderly can be 55~75 times/min, and newborns can be fast to 120~140 times/min.

Normally, a person's heart rate slows down when he or she is quiet or sleeping and accelerates when he or she is exercising or excited. Under the effect of some medicines or neurohumoral factors, heart rates can also speed up or slow down. Continuous pulse monitoring has great positive significance for the real-time control of people's physical state. At present, the wearable intelligent hardware used for monitoring heart rate mainly uses photoelectric signals; for example, Apple Watch adopts optical technology of photoplethysmography (PPG) using LED green light and infrared light. There are two other types of light sensors that measure heart rates by blood flow detection. Pulse can also be measured by pressure sensors or ECG sensors.

3. **Breath**

Breathing is the activity of the respiratory tract and lungs. The human body inhales oxygen and exhales carbon dioxide through breathing. It is one of the critical life functions and cannot stop for a moment. It is also the necessary process for gas exchange between the internal and external environment of the human body. Normal people have smooth respiratory rhythm and appropriate depth.

Normal smooth breathing is 12–20 times/min for adults and 30–40 times/min for children. The breathing times of children decrease as they grow and will gradually reach an adult level. The ratio of breathing number and of pulses is 1:4. The breathing of normal people could accelerate to more than 20 times/min when excitement, exercising, eating and temperature rises. Breathing could slow down when they are sleeping. At present, the mainstream method of monitoring is to use pressure or motion sensors. Breathing-related data can be obtained by collecting and processing chest/abdomen motion signals.

4. **Blood pressure**

Blood pressure means the lateral pressure of blood in a vessel against a unit area of the blood vessel wall. It is one of the critical indicators to measure cardiovascular function, which generally refers to arterial blood pressure. When the ventricle contracts, the highest pressure in the artery is called systolic pressure; when the ventricle is diastolic, the lowest pressure in the artery is called diastolic pressure. The difference between systolic and diastolic pressure is pulse pressure. The location of blood pressure measurement is commonly the brachial artery of the upper arm. Intermittent measurement methods such as Korotkoff sound are often adopted to measure blood pressure in clinical practice. This method is hard to conduct in real-time monitoring, and it only obtains the blood pressure value at a certain moment. Blood pressure measurement is affected by various factors, such as emotional excitement, stress, exercise, etc. Although most blood pressure values measured by wristbands and watches are not completely accurate, they can basically reflect the correct trend of blood pressure change. In addition, observing the continuous blood pressure trend is of great practical significance. Currently, blood pressure tracking measurement of wearable equipment based on photoplethysmography (PPG) and electrocardiography (ECG) is combined with the wrist watch, which has been listed.

In addition to the 4 basic vital signs of body temperature, pulse, breath and blood pressure, there are other vital signs, such as blood glucose and oxygen saturation. Reports show that some research has taken the principle of "the heavier the psychological stress, the higher the level of blood oxygen". Through measuring the facial oxygen content in the blood, an intelligent camera has been developed that can look for highly nervous people in the crowd. This provides an effective means to identify suicide attackers.

The core of vital sign monitoring lies in sensors, such as temperature and humidity sensors, pressure sensors, and displacement sensors. Since 2013, the bloom of intelligent wearable equipment, including intelligent wristbands and intelligent watches has emerged quietly. Apple Watch and Google Glass are all wearable equipment. Wearable vital sign monitoring equipment is considered the next huge growth point in the technology industry. Wearable equipment with multiple sensors can not only track users' fitness but also offer geographic location-related functions, such as maps, shopping, and even financial services. Wearable equipment is not only hardware equipment but can also realize powerful functions through software support, data interaction and cloud interaction. The built-in photoelectric pulse sensor can grasp the blood flow when exercising and measure the pulse through the capillaries on the wrist surface. At the same time, an accelerometer is integrated that can measure the pulse data and obtain the walking data just by wearing it on the wrist.

The sports wristband monitors energy consumption through the built-in simple pede-counting sensor. People are expected to know the more complex health conditions by wearable equipment, such as the blood glucose index and stroke risks. Better warning effects can be achieved through electronic wristbands with positioning and sensing functions.

The sensors of wearable intelligent equipment can be divided into.

(1) Motion sensors

This mainly includes accelerometers, gyroscopes, magnetometers and pressure sensors (calculating altitude by measuring atmospheric pressure); these sensors are used for motion monitoring, navigation, entertainment, human–computer interaction, etc.

(2) Biosensor

This mainly includes blood glucose sensors, blood pressure sensors, ECG sensors, electromyography sensors, body temperature sensors and brain wave sensors.

(3) Environmental sensors

This mainly includes temperature and humidity sensors, gas sensors, ultraviolet sensors, ambient light sensors, particulate matter sensors, air pressure sensors, microphones, etc.

Prison vital signs monitoring applications generally adopt electronic wristbands combined with sensors to collect human vital sign data. These sensing data must undergo credible analysis, filter out interference from the outside environment, combine them with other information sources (such as people's motion state information) and business scenarios, and finally can be mined and utilized.

## 6.2 Development of Biometric Technology

Biometric technology conducts personal identification through the combination of computers with optics, acoustics, biosensors, biostatistics principles and other high-tech means, using the human body's inherent physiological characteristics (such as fingerprint, finger vein, face, iris, etc.) and behavioral characteristics (such as handwriting, voice, gait, etc.). Human biological characteristics are usually unique, with features of measurable, autoidentifiable and verifiable, hereditary or lifelong invariables. Thus, biometric technology has more advantages than traditional identification technology. Biometric technology has experienced a development process from external characteristic identification to internal characteristic identification, and this reflects the three levels of prison vital sign monitoring applications.

The first level: based on identification needs.

The construction of the Prison IOT vital signs monitoring system requires each object (equipment or human) in the target system to be configured with a unique corresponding identification code. This is the basic technical requirement for the architecture IOT sensing layer. To meet the object recognition requirements, it commonly adopts the following two methods.

1. Taking the shape characteristics of target objects as the recognition basis. The unique characteristics of an individual object (equipment or human) are collected as an identification symbol, used to determine the real identity of the object rapidly and accurately. Fingerprint, palmprint, vein, iris and other biometric characteristics belong to this category.
2. Taking the RFID electronic tag as the carrier, through one-to-one correspondence, binding with the unique object. This achieves the requirement of object recognition by the identification of RFID.

Taking the shape characteristics as the recognition basis, the collection process is easily affected by environmental factors, and its recognition efficiency is low due to individual differences, which largely limits its application in a large area. Taking an RFID electronic tag as an identification marker may easily lead to confusion within the identification system if the carrier is changed. An anti-detachment wristband is adopted to realize the uniqueness of RFID tag identification.

The second level: based on location needs.

After the problem of "whom" is solved by intelligent means, the following problem needs to be solved in Prison IOT vital signs monitoring system construction, including the need of "where".

With the basic technology of "recognition", to realize the collection of location information, in addition to taking traditional video analysis as the recognition basis, the RFID electronic tag combined with sensors should be a more appropriate choice to realize positional recognition.

(1) Realize real-time monitoring of locations

Personnel, as one of the core elements of IOT real-time location information, should be correctly displayed in the IOT. Using RFID technology to "tag" people can achieve real-time monitoring of the location of specific people in specific areas.

(2) Realize monitoring based on location state

Through the combination of displacement, velocity/acceleration sensors and RFID electronic tags, real-time monitoring for the moving speed/moving direction of the monitored object can be achieved. The real-time collected movement status data can also accurately reflect whether the area where the object is located complies with the authorization requirements, and it can also be expanded to some functions of counting the number of prisoners in certain areas of the prison, trajectory query, and cross boundary alarm.

The third level: based on behavior needs.

After the "whom" and "where" problems are addressed, the next step in Prison IOT construction is solving the deeper information of "what is being done", that is, the behavioral attributes of the object.

Behavior collection and analysis is a complicated project that requires a large amount of credible data to support it. From the perspective of front-end perception, based on completing the real-time location monitoring of people, it is suggested to grasp the real-time state information of the object—from the external movement status information to the vital signs information like body temperature, heart rate, blood pressure, etc. Eventually, it will evolve to the acquisition of a deeper level of psychological attributes, information on emotion and stress.

(1) Real-time behavior status monitoring

The behavior status data is mainly gathered by location, acceleration, and altitude sensors against the perception of human behavior states, including states of resting or walking, moving or falling, or whether they are in a certain state for a long time to analyze and judge human behaviors.

(2) Real-time vital signs monitoring

The collection of vital sign data usually uses temperature sensors to detect body temperature, pressure sensors to detect pulse/blood pressure, chemical sensors to detect sweat composition (such as alcohol content), and microelectric signal sensors to monitor heart/brain/electromyography. Through the monitoring of the vital signs of pulse, body temperature and blood pressure, it is possible to further analyze the behavior characteristics of the supervised object.

(3) Interaction with external information

If state or vital signs are monitored, this has limitations. To obtain perception data with more application value, integration with more external information (such as environment information). By combining velocity and acceleration sensors with real-time vital sign data monitoring, an object's movement behavior data (static, movement, slip, etc.) can be obtained, attached to basic conditions for the credibility of real-time vital sign data and combined with objective external environment and time features (for example, distinguishing sleeping at night or coinciding). This constitutes the 4 W's (Who, Where, What, When). This can form a human-centered perception data file by comprehensive analysis and conducting a deeper level of data mining and behavior analysis based on perception data.

## 6.3 The Evolution of Vital Signs Monitoring Methods

IOT technology addressed perception issues through sensors and the use of the network to achieve remote monitoring, similar to human eyes, ears and skin. One of the reasons for this revolutionary change is the fundamental change in monitoring methods. Some information originally acquired by artificial means can be automatically acquired with the help of technological means. For instance, the location information of personnel can be obtained in real time with the assistance of electronic tags, and some parts of the behavioral data of people can also be obtained by intelligent video analysis.

Vital sign monitoring methods will become more convenient in the future. For example, the collection method for heart rate currently adopts the advanced means of combining wearable intelligent hardware with sensors. In certain application scenarios, the noncontact monitoring method of "no need to wear" to collect continuous and real-time vital sign data could be a more acceptable solution.

IOT technology has gradually matured, and there are new tools and means for positioning management, cross-boundary management and behavior recognition of criminals in prisons. Placing electronic wristbands or electronic foot loops on prison inmates causes prisoners to become one of the prisons IOT information points. Through receivers sending the movement speed, direction, height, vital signs (body temperature, pulse) and other information of the wearer to the background by network. The effective management of prisoners can be realized through information integration and analysis.

## 6.4 Design of System Overall Architecture

The prison vital signs monitoring system, as the foundation of the perceptual layer architecture, is based on the combination of RFID radio frequency identification and sensor technology. In addition, it takes the supervised object as critical to achieve centralized perception of identity, vital signs and external environment. Different types of sensors conduct reliable and safe intelligent transmission by building a unified prison IOT sensing grid as the transmission layer. Through prison IOT middleware technology to realize the centralized management of prison sensing information, all sensing data can be shared with other prison security and business platforms. Big data application scenarios could incorporate prison business needs to select appropriate sensing information for data mining. Design of the architecture is needed to meet the following technical requirements:

◎ Safety. The whole network should comply with the prison's unified and dedicated radio frequency protocol to achieve isolation between the radio communication network and the wired communication network. Avoid the possibility of accessing the prison's internal data network through a radio frequency communication network (physical isolation or gatekeeper). Transmission should be limited to the relative position, vital signs and environmental sensing information, which requires encryption.

◎ Expandability. The whole sensing network should comply with the principle of dynamic expandability. Systems can support the access of various subsequent types of sensing tags.

◎ Relatively open. All the sensing data related to the object's identity, location and vital signs should be owned by prisons, which is a valuable asset of the prison. In addition, after ensuring privacy and safety, these data can be opened and used for other related business systems inside the prison.

◎ Maintainability. Systems offer sensing-end equipment and transmission layer equipment with automatic maintenance functions to decrease the maintenance cost of systems by means of unified platform management for all equipment.

The overall architecture of the prison vital signs monitoring system as shown on the Fig. 6.1.

The mainsteaming of a prison vital signs monitoring system is based on the object's vital signs sensing data. The "comprehensive perceptions, collaborative linkage, sharing and exchange" of the object's sensing data is realized through the prison IOT middleware platform and prison IOT comprehensive management platform.

The main components of the Prison vital signs monitoring system include the following:

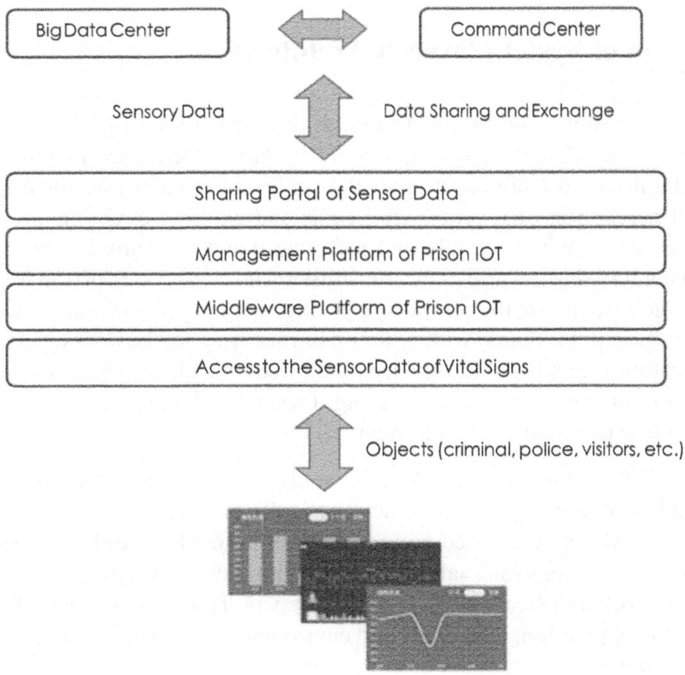


**Fig. 6.1** The overall architecture of the prison vital signs monitoring system

(1) Wristband sensor tag

Vital sign data monitoring generally uses wearable electronic wristbands (or cards), and the corresponding vital sign sensors are integrated inside by wearing tags on criminals' wrists. The basic vital sign data of tags can be recognized by radio signals received and analyzed.

(2) Noncontact vital sign sensors

Noncontact vital sign sensors are advanced and convenient IOT sensing technologies. It is different from traditional technology with its feature focus on "noncontact" but which can perceive the basic vital signs of the object. For example, under the condition of not having contact with a criminal's body, by installing special sensors on the prisoner's bed, the heart rate, breath, movement, location, motion trajectory, microexpression, sleep state and other criminal data can be continuously and dynamically collected and analyzed in real time. This can help examine the change in the discipline of criminals' emotional state and behavioral mode in stressful situations.

(3) Prison IOT middleware

The prison IOT middleware system is a service-oriented software framework that provides service to the upper application system by a standard service interface. Cross-heterogeneous systems are utilized to realize the unified management of personnel positioning, video surveillance, entrance guards and boundaries through the unified access, unified analysis and classification processing of sensing data. The key task of the system is to carry out redundant filtering, integration and release against the collected control data of personnel positioning, video surveillance, entrance guard and boundary to perform standardized transformation for various information formats and protocols.

Another task is to provide an upper application system or data management system with valuable data and, in the meanwhile, shield the difference between various RFID readers, realize transparent integration between various parts, facilitate the development of application systems and play an intermediary role between RFID readers and application systems.

(4) Prison IOT comprehensive management platform

A prison IOT comprehensive management platform is the system to comprehensively control the IOT sensing data, which is managed by a unified interface. The management platform can perform continuous real-time monitoring and comparative analysis against data to make appropriate early warnings in abnormal conditions. At the same time, data can be provided for various business systems through an open API interface to achieve the sharing of sensing data.

## 6.5 Ideas for Designing System Functions

The following is the design of prison IOT vital sign monitoring system functions:

(1) Real-time monitoring of heart rates and breaths

Real-time monitoring of the criminal's vital sign signals by sensors mainly includes real-time heart rates and breaths. There is a need to focus on monitoring criminals who are ill or who may attempt self-harming.

(2) Prediction on sleep and mental states

This makes a psychological trend analysis of monitored personnel through long-term monitoring of the criminal's vital signs and sleep status to predict the criminal's behaviors. Monitoring data can directly incorporate psychological tests, psychological health education, psychological crisis intervention and individual education.

(3) Alarm of abnormal situations

According to preset rules, alarm in real-time under abnormal situations that do not conform to the rules:

① Alarm of leaving bed and not returning. Alarm when the criminals get out of their bed (such as to go to the bathroom) exceeding a certain period and do not return.

② Alarm of abnormal vital signs. Alarms will be activated when a criminal has abnormal vital signs, such as sudden abnormal heart rate monitoring.

(4) Single police life monitoring and supervisory security assurance

Establish a supervisory security assurance information chain based on a single police life monitoring system, which has great value for implementation of positioning rescue and identification confirmation in emergency locations. Through vital signs sensors, it can specify preparation for emergency rescue, accurately dispatch supervisory force and resources, and comprehensively improve the prison safety assurance ability.

(5) Psychological assessment of criminals

Through the full-period multidimensional acquisition of criminals' physical (brain electricity, heart rates, breath, skin electricity), biochemical (cortisol) and behavior indicators (microexpressions, actions, behavioral performance), one can build a criminals' multimodel physiology-biochemical-behavioral database under normal, stress, task states. This will provide critical data for prisons with better accuracy in the identification and correct judgment of criminals, provide critical data for establishing qualitative and quantitative scientific psychological assessment risk indicators and provide critical data for individualized education correction.

## 6.6 Problems and Prospects

The accuracy of vital sign monitoring data is affected by the accuracy of the sensor. The standardized wearing by users and the reliability of the transmission network will reflect on the data mining and utilization. Based on different application scenarios, improving the accuracy of the collected vital sign data is a technical challenge. One set of credible assessment models must be established to efficiently filter out "noise" data in the link of data mining and utilization to reduce the errors caused by external interference.

The integration of vital sign monitoring technology and prison personnel positioning technology can realize risk evaluation of criminals' dangerous behaviors based on the prison's personnel locations. Analyzing the acquired information of each criminal's real-time heart rates, breath rates and sudden abnormal physiological status, combined with data and rules provided by the prison management business system, active monitoring, automatic measuring and appropriate early warning against suspicious, illegal, irregular and other abnormal behaviors (such as depressed behaviors) can be realized. It provides critical data for prisons with precise identification and accurate judgment of criminal risk as well as qualitative and quantitative establishment of scientific risk indicators.

# Chapter 7
# Prison IOT Application—Local Area Positioning System

According to the concept of wireless positioning in prisons, currently this is used mainly to achieve indoor positioning accuracy of a 0.5~5 m, error-free distinction between an indoor specific room and area, error-free judgment at the floor, and an error- free difference between indoor and outdoor and outdoor positioning accuracy of 5~10 m. The development of wireless positioning systems in prisons could revolve around the abovementioned accuracy to design application scenarios. With the continuous development of wireless positioning technology, positioning accuracy will continue to improve.

Influenced by various factors, at present, the demand for accurate positioning in prisons has been overinterpreted. High hopes are placed on pilot wireless positioning applications in prisons to achieve scientific effects. As a result, there are doubts about wireless positioning technology. For the application of RFID electronic wristband technology in prisons, the first problem that needs to be solved is to informatize people (criminals, police, visitors, etc.), that is, through the electronic tag (wristband), people are to be converted into an information node, which could be connected into the prison network. Meanwhile, through this information node, the relevant data could be merged (such as the implementation of punishment correction) to realize a "single number reach". However, according to the analysis of the wireless positioning technology applicable to the prison in the previous chapters, the condition of current technology, the accuracy of active RFID, WiFi technology, and ZigBee technology is approximately 2~5 m. This makes it hard to realize the accuracy required in prisons, even by improving the positioning algorithm or by adding auxiliary reference nodes. Therefore, it is still difficult to compare UWB technology with a positioning accuracy of 10~30 cm.

This chapter and the next will discuss the local area positioning of the prison based on active RFID technology and the relatively accurate positioning of the prison based on UWB technology.



P. Sun, *Smart Prisons*,
https://doi.org/10.1007/978-981-16-9657-2_7

## 7.1 System Overview

Since information construction must be "effective, practical, and real", the issues of prison management corresponding to the complementation of wireless positioning projects in prisons are as follows:

(1) It is impossible to accurately know the number and location of prisoners in real time

Accurately knowing the number and location of prisoners is the primary factor to ensure the safety of surveillance sites. Prison police need to know the location of the criminals in real time if they are out of control, entering and exiting specific areas illegally or illegally approach important facilities. The traditional method of confirming the number and location of people mainly relies on a continuous prison police count. However, criminals are moving constantly in different areas, and because of insufficient police, individual misreporting, missed reports, concealed reports, etc. Relying on the police count and reporting to the next superior administration, the summarizing speed is slow and does not provide efficiency.

(2) It is impossible to strictly identify the identity of criminals without making mistakes

The identification of criminals in prison is still relatively traditional, and manual identification makes it difficult to ensure that there is no error.

(3) It is difficult for criminals to go out temporarily and to monitor the entrance and exit of visitors

When criminals go out for medical treatment, arraignment, transfer to another prison, etc., they need to leave the prison, and the risk of escape is relatively high. At present, supervision mainly relies on police and guns and lacks high-tech means for assistance. In addition, for foreign vehicles and construction and maintenance personnel entering and leaving the prison, meeting lawyers and family members also requires positioning management.

The construction requirements of the prison wireless positioning project can be classified into the following two points:

(1) Realize the positioning management of the police, criminals and outsiders. Real-time positioning, automatic roll call, video linkage, track playback, alarm management, query statistics and other auxiliary functions can be realized by wearing electronic tags (electronic wristbands).
(2) Realize the positioning management of important assets such as special items (such as keys) and dangerous labor tools. By installing electronic tags on dangerous labor tools and important assets, real-time positioning and automatic counting can be realized in asset management, labor tool distribution and homing.

## 7.2 Overall System Architecture Design

The architecture design of the prison wireless positioning project needs to meet the following technical requirements:

- Integrity. The overall design of the system should be unified and standardized, and the design of functional modules must be clear, reasonable, and able to effectively integrate with prison management and command platforms to achieve prison integration and automatic management.
- Advancement. The system design plan should conform to the mainstream technology and ideas of the current information system development trend, and the technology should be mature.
- Practicality. The system design should fully consider the actual work characteristics of the prison and the actual situation of the existing software and hardware systems so that the system is strongly practical.
- Reliability. The system design must fully consider stability and reliability to ensure the correctness under normal and extreme conditions, avoid affecting the normal operation of the entire system due to the failure of a single point, and ensure that the system functions to the utmost extent.
- Security. System design must consider not only information security but also the safety of the human body.
- Scalability. When the system capacity changes, it should be able to ensure a reasonable response time and handling capability of the system through various horizontal and vertical expansions.
- Accessibility. The system should provide a unified style, attractive user interface and convenient and practical use methods so that users can master the operation in the shortest time.
- Maintainability. System software and hardware equipment should be able to be easily and conveniently managed, modified and upgraded.

The evaluation criteria of the prison wireless positioning project can be referred to as follows:

- Positioning method and algorithm selection.
- Positioning response time and error size.
- Anti-collision capability (tag reading density).
- Anti-interference ability.
- Tag battery life and whether it supports the hibernation function.
- Tag size, shape, IP protection level, etc.
- Information security capabilities.
- The difficulty of deployment and the length of the implementation period.
- Reuse of existing infrastructure.
- Return On Investment (ROI).

The overall architecture of the local area positioning system based on active RFID is shown in Fig. 7.1.

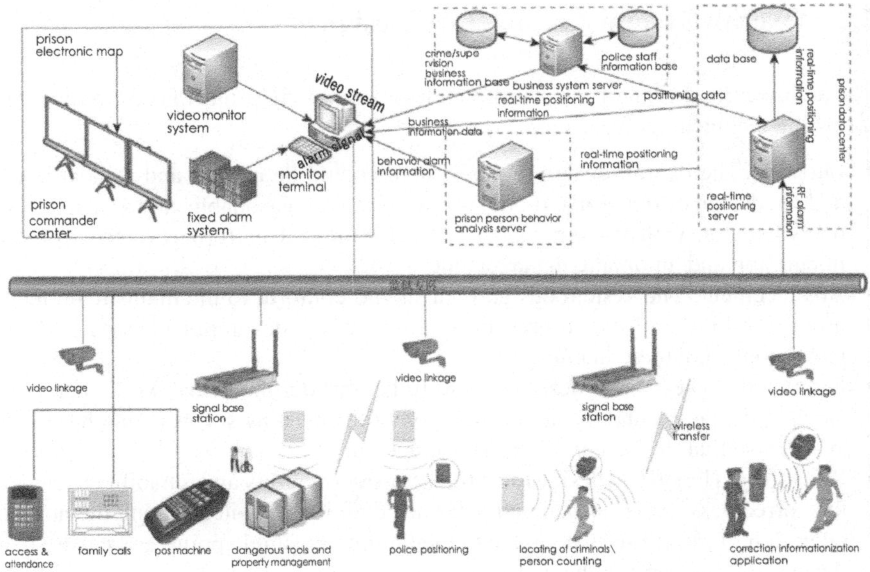


Fig. 7.1 Local area positioning architecture diagram based on active RFID

The real-time positioning server calculates the relative real-time position of people and objects by receiving radio frequency signals emitted by the RF positioning identification card worn on policemen, the electronic wristband worn by criminals and the tags of the asset and labor tools; the prison personnel behavior analysis server analyzes peoples behavior based on location; the business system server can provide basic data about criminals and prison police and supervision business data; combined with electronic maps, real-time location information and movement tracks of police, and criminals and visitors can be displayed in the prison GIS in real time.

The main components of the local area positioning system based on active RFID include:

(1) Anti-disassemble electronic wristband (for criminals)

The radio frequency positioning electronic wristband is a radio frequency transmitting device worn on the criminal's wrist, which identifies the criminal by sending a unique number at regular intervals. In addition, the electronic wristband transmits radio frequency signals to analyze the location of the electronic wristband and report the status of the wristband, such as battery status, whether it has been disassembled, etc.

Generally, multiple frequencies can be integrated in the electronic wristband, corresponding to a variety of applications, such as integration of 13.56 MHz for access control and attendance, family calls and electronic consumer card swiping. The latest applications also include integrated vital sign monitoring systems.

(2) Radio frequency positioning identification card (for police and visitors)

The radio frequency location identification card is a radio frequency transmitting device worn by police and visitors, and it is used to identify the police and the individual visitor by sending unique numbers at regular intervals. In addition, a hidden alarm button is provided on the card. When an emergency occurs, one can press the button (or pull the lanyard) to make the identification card send an alarm signal. The alarm signal is received by the radio frequency signal analyzer and transmitted to the real-time positioning server.

(3) Real-time positioning server

Through receiving the radio frequency positioning signal and the alarm signal, according to the content of the signal data package, the real-time positioning server calculates the real-time position of each radio frequency positioning tag and sends the alarm signal to the monitoring terminal of the command center immediately. The real-time location server can also provide real-time location storage, playback, behavior analysis, early warning and other functions.

(4) Prison staff behavior analysis server

The prison personnel behavior analysis server monitors violations, dangerous behaviors, or behaviors that reflect specific psychological characteristics in real time and can compare and analyze historical data, which can realize timely alarms on illegal gatherings, gangs, suicides, or police violations to prevent and promptly handle accidents in the prison.

(5) Prison electronic map (GIS)

The prison electronic map receives various information, such as real-time location and alarm, from the real-time positioning server, displays the location and movement track of the police and criminals in real time on the electronic map and presents various warning information to the police in the form of sound and images.

The electronic map platform can also provide interfaces with video surveillance and alarm systems to realize the function of video and alarm linkage. The police can perform rich management, monitoring and emergency response functions on the GIS platform, such as scene roaming, personnel retrieval, track playback, video review and alarm handling.

## 7.3 Design Ideas of System Function

Prison wireless positioning applications have changed the goal of prison information construction from passive security, such as traditional perimeter access control to active security, that is, with a "person" in the prison as the core. Now, this information is converted into points through electronic tags and connected to the perceiving terminal of the Internet of Things in prison. Prison police can grasp the number and location information of criminals in each controlled area in the prison in real time, including the information of the surrounding police on duty, and can realize the alarm function for various emergencies. When there is an emergency, the police can quickly locate the relevant criminals, the police on duty, and can quickly view the historical movement track.

1. **Basic functions**

(1) Automatic number counting

According to the needs of the supervision business and the actual site environment, the entire supervision site is divided into different areas, such as prisons, hospitals, canteens, teaching buildings, etc., for the automation of counting the number of criminals.

(2) Real-time positioning of personnel

In design and application, the real-time positioning of personnel can consider "highlight management and control requirements and weaken positioning functions". Unless there are behaviors such as escape, rIOTs, illegal entry and exit of specific areas, illegal approach to key facilities, etc., the system will immediately locate the area of the alarm and historical track information and can provide on-site video images for further analysis and processing by the police. Where regular positioning data is displayed normally, this is not necessarily required.

(3) Out-of-prison management and monitoring

When there is a need to leave prison temporarily due to medical treatment, the police can connect the police mobile handset to the criminal's electronic wristband. Once the criminal has left the police away from the specified distance or the wristband is damaged, the police can be alerted, and the alarm information will be transmitted back to the prison command center.

(4) Real-time alarm

1 Positioning tag (wristband) signal loss alarm

When too many radio frequency tags in a certain area cause conflicts or the tags leave the signal coverage area, the positioning tag (wristband) signal will be lost, and the system will alarm immediately.

2 Alarm for damage of positioning wristband

The electronic wristband itself should have an anti-damage design. If there is an event that destroys the internal circuit, such as cutting or forceable removal, the system will immediately send an alarm.

3 Alarm for noncompliance with predetermined rules

- Alarm of retention in the area. Alarm for personnel staying overtime in a certain area (overtime staying can be flexibly set).
- Alarms that do not enter the designated area on time. If personnel do not reach the designated area within the specified time, the alarm will sound.
- Alarm for entering an illegal area. If the criminals enter a prohibited area, or violate rules by visiting other areas, an alarm will be issued.
- Alarm for leaving a certain area without authorization. If the criminal leaves an area without authorization, the alarm will be issued.
- Alarm for track deviation. If the track deviates from the preset route, an alarm will be issued.
- Alarm for unauthorized gathering. For unauthorized times and areas, the system can be set for the maximum allowable number of persons to be gathered, and once there are more than the set number of criminals gathered, an alarm will be issued. This function can also be extended to issue the alarm for any tags in a gathering, such as a setting that persons in the same case cannot meet or two suspects fighting each other cannot be together.

4 Alarm call

Criminals, police, and visitors can quickly call the alarm by pressing the alarm button on the wristband or card under special circumstances.

5 Alarm for low battery

When the battery of the positioning tag (wristband) is low, it can send out a low-power alarm signal to remind staff to replace the battery.

(5) Query statistics

The query statistics function provides statistics on positioning alarms and area entry and exit in the system and provides quantitative data for management. You can send a query based on region, time, personnel category, and specific individuals and you can also query in any combination.

- Regional query statistics. All relevant personnel or regional management personnel can perform statistics on their affiliated management areas, such as querying and counting the number of people who should arrive, the actual number of people, the number of abnormalities or the number of alarms in the area within a certain period of time (administrators with high-level authority can perform management statistics for their subordinate areas).

- Query statistics of different personnel categories. One can query the number of criminals, police, and temporary foreign personnel (such as family members, lawyers, prosecutors, equipment maintenance personnel, etc.) and the area in which they are located.
- Time query statistics. One can count the entry and exit of personnel, alarms, etc. within a certain period, such as counting the number of people going out for medical treatment within a month and counting the alarm situation within a week.
- Query of personal tracking. One can query the track of a criminal in a day and can link back to check the video of entering and exiting the relevant area at that time.

(6) Seamless integration with the command platform

Through seamless integration with the command platform, one can achieve coordinated linkage with video, access control, perimeter, and business systems to meet active and intelligent security requirements.

(7) System management

1 Location tag management

- Automatic detection of positioning tags

  The system automatically detects the status of all locating tags in use, and if any abnormal situations are found, such as sudden or long-term failure to receive a locating tag signal, the system will alert the police to check whether the locating tag is damaged.

- Add, delete, modify and check functions of positioning tags

After issuing an electronic wristband to a newly admitted criminal, the police can input the electronic wristband number in the system, associate the detailed information of the criminal at the same time, and delete it when leaving the prison.

Temporary electronic tags used by visitors can be quickly relevant.

2 Alarm rule management

According to actual demand, you can customize the activation time period for the intrusion of the guard zone and the escape of criminals to issue alarms. You can even customize the guard zone range and boundary, as well as alarms for individual criminals to go out of the zone. The system should be flexibly set up to facilitate the management of the police and reduce the occurrence of false alarms.

3 Personnel access authorization

Perform temporary authorization on designated criminals and allow them to enter and exit the warning area or controlled area. Authorization must include criminal information, authorized time, authorized area range, responsible police information, etc.

Temporary electronic tags used by visitors need to be authorized for the range to enter and exit.

4 User authority management

The system operation authority can be set according to the needs of different users, such as adding, querying, modifying, and deleting login user information, criminal information, police information, and electronic tag (wristband) information.

2. **Extended functions**

(1) Important asset management

Install electronic tags with displacement sensors on important assets. Once the assets move, an alarm will be issued. The tag of the asset and its owner can be bound. If the system detects that two tags are not present at the same time when entering and exiting the door, an alarm can be issued.

(2) Firearms and police weapons management

Firearms and police weapons can also be equipped with an anti-removal electronic tag with a displacement sensor. Once a gun is forcibly removed from the gun holder, the electronic tag immediately sends a special ID to the receiver, and the receiver receives the signal and immediately determines whether to issue an alarm. In actual use, guns and ammunition can be stored separately, and different authorized electronic tags can be used for control to ensure safety.

(3) Management of dangerous labor tools

The standardization of the management of dangerous labor tools directly affects the security and stability of prisons. Through the integration of labor tools and electronic tags, labor tool registration, issuance, return and other procedures can be implemented simultaneously in the system to truly "clear the number and clear the situation" and realize all-weather, full-coverage dynamic management.

(4) Integrated vital signs monitoring

Vital sign monitoring can be integrated into the electronic wristband, and patients or criminals who may attempt suicide or self-harm can be monitored.

## 7.4 Problems and Prospects

Regardless of any positioning technology used, special attention must be paid to the three most important technical parameters: positioning accuracy, communication distance and battery life. The best solution must meet high positioning accuracy, longer communication distance and longer battery life at the same time. The smaller the size, the better, but these points are contradictory and restrictive among each other under current technical conditions. From a technical point of view, the higher the positioning accuracy, the better the user experience will be. However, related issues such as the corresponding cost and implementation complexity of the solution will also be clearly exposed. In fact, in most current prison application scenarios, local area positioning can already achieve good application effects, which can fully meet the needs of users without deliberately pursuing high positioning accuracy.

According to the needs of local area positioning, the access control system can actually be the simplest example of local area positioning. An example would be by combining with other perimeter prevention systems, such as setting up infrared fences at the exit, when it is confirmed that no electronic tag cross-border incident has occurred, and this can determine that the tag is in this room. However, the smallest unit of the positioning accuracy of the access control system is limited by the specific indoor structure of the location.

When installing prison wireless positioning applications, attention should be given to wireless network surveys and design in advance. At present, most of the problems are focused on the inability to accurately determine the number of installation and installation methods of related equipment (such as tag readers) before installing wireless positioning applications. The survey and analysis of the wireless network may help to achieve the correct bandwidth, speed, and overall signal coverage effect, help to determine the type and quantity of positioning AP, antennas and other components and help to determine the ideal location of the positioning AP and bearing angle of antennas and other engineering parameters. This helps to quantitatively analyze the radio frequency and noise source and to monitor the connection speed, retery rate, and data package loss of wireless nodes.

For applications that use RSSI signal strength and positioning technology, pretraining is required before use (to precollect and record the received signal strength of RSSI values from various grided locations of the prepositioning area and then determine the location based on signal distance or distribution features), so it is necessary to conduct coverage surveys for wireless signals.

When installing a prison wireless positioning system, the following points need to be considered:

(1) The number of people per unit area in the prison is large and dense. The implementation of positioning practice simultaneously is a major challenge. It is necessary to fully consider the real-time and accuracy of the positioning system, as well as the anti-collision and anti-interference capabilities of the positioning base station to prevent the missed reading of the positioning tag signal.

(2) Most prisons have a large number of buildings. To achieve full positioning coverage, much positioning equipment needs to be installed. In most older prisons, since strong and weak electrical pipelines have been laid as it is easy to install, easy to manage and maintain, etc., corresponding requirements must be put forward.
(3) The drift of wireless positioning signals can generally be corrected by methods such as auxiliary positioning and software correction.
(4) The prison wireless positioning system must be integrated with various prison monitoring systems (command platform) to realize early warning linkages.
(5) Considering the privacy and security of prison data, the prison wireless positioning system should be considered information security.

# Chapter 8
# Prison IOT Application—Relatively Accurate Positioning System

Accurate positioning is a relative concept. It is difficult to evaluate how many meters of accuracy can be regarded as regional positioning and how many meters can be regarded as accurate positioning, particularly in today's rapid development of indoor positioning technology. From the needs of prison radio positioning applications, whether the accuracy of personnel positioning meets the requirements of "Nanorobots performing tasks of precisely positioning body areas and delivering drugs", this question may not be answered immediately.

UWS technology and CSS technology can achieve dozens of centimeters-level positioning. This chapter will discuss relatively accurate positioning based on UWB technology.

## 8.1 System Overview

Ultraw-wide band (UWB) radio technology is an advanced wireless communication technology using bandwidths above 1 GHz and does not need a carrier wave. The information carrier inside the UWB radio is impulse radio (IR). IR refers to wireless technology that adopts shock pulses (ultrashort pulses) as information carriers. The feature of this technology is to obtain an ultrawide bandwidth for data transmission by modulating ultranarrow (often less than 1 ns) pulse signals. UWB uses a radio wave bandwidth of several gigahertz waves, which is hundreds of times higher than a wireless LAN of approximately 20 MHz. Its communication speed can reach in excess of hundreds of megabits per second.

UWB has experienced the following three phases:

The first phase: Before 1989, UWB signals were obtained mainly through transmitting ultrashort pulses. This technology is commonly used for radio fields and uses the term "pulse radio," which belongs to carrier-free technology.



P. Sun, *Smart Prisons*,
https://doi.org/10.1007/978-981-16-9657-2_8

The second phase: In 1989, Defense Advanced Research Projects Agency (DARPA) initially used the term "UWB" and stated that if the absolute bandwidth of a signal at 20 dB attenuation is greater than 5 GHz or the relative bandwidth is greater than 25%, then the signal is a UWB signal.

The third phase: To facilitate and regulate the development of UWB technology, in April 2002, the FCC released the initial regulation of UWB radio devices and refined UWB. According to UWB, the bandwidth of the UWB signal should be larger than or equal to 500 MHz, or its relative bandwidth is greater than 20%.

UWB radio communication technology was mainly limited to research on government and military projects before the 1990s. In February 2002, The United States Federal Communications Commission (FCC) allocated a total bandwidth of 7.5 GHz from 3.1 to 10.6 GHz for commercial UWB communications. The IEEE 802 Committee also took UWB as a candidate for the fundamental technology of the Personal Area Network (PAN). In March 2007, ISO officially passed MB based on the OFDM standard submitted by WiMedia Alliance and officially became the first international standard of UWB technology. UWB technology has been recognized as the revolutionary progress of radio technology. Its huge potential has broad application prospects in wireless communication, radar tracking, and accurate positioning.

The UWB frequency range has the following advantages for positioning technology:

(1) Good anti-interference performance. As the UWB radio transmits a single-period pulse with a very short duration and a very low duty cycle, UWB spread spectrum processing is dependent on the duty cycle of pulses and the number of pulses used to send each signal. The duty cycle of pulses is generally 0.01–0.001, with higher processing gain than other spread spectrum systems, strong anti-interference and its anti-interference processing gain is 50 dB. In addition, the multipath resolution capability is strong. UWB has very high resolution due to its extremely high working frequency and extremely low duty cycle. Narrow pulse multipath signals do not easily overlap in time, so it is easy to separate multipath components to use the energy of the transmitted signal efficiently.

(2) Extremely Low energy consumption. UWB systems use intermittent pulse to transmit data, the pulse duration is short and generally between 0.20–1.5 ns. Its duty factor is low, and the power consumption of the system can be very low, at approximately 1/100 of the power required by traditional mobile phones and 1/20 of the power required by Bluetooth devices.

(3) Simple structure. Due to the use of pulses for signal modulation and transmission, the antenna is excited directly by small types of pulses, without the upper conversion required by traditional transceivers and does not require power amplifiers and mixers. Thus, UWB can use cheap broadband transmitters.

(4) High safety. The power spectral density of UWB signals is lower than that of natural electronic noise. It is very difficult to detect pulse signals from electronic noise. After the pulse parameters are pseudorandomized by coding, the detection of the pulse will be more difficult.

(5) High positioning accuracy. The shock pulse has high positioning accuracy, which can provide centimeter-level positioning accuracy indoors and outdoors.

With present markets, mature commercial UWB products can reach positioning accuracy of approximately 10~30 cm. The current principle of accurate positioning is related to the radio measuring distance principles. The combination of Time Difference Of Arrival (TDOA) and angle of arrival (AOA) is the most widely used location method in UWB location technology. Radio waves determine their positions by measuring the path from the transmitter to the receiver; only the straight path is true, and other reflected paths are faults. The reflection will distort the straight path signal when positioning by the traditional RF method, which makes accurate positioning more difficult. UWB technology can recognize the direct path signals from the reflected signals, which makes accurate positioning simpler. Since the shock pulse has a high level of positioning accuracy, adopting UWB radio communication makes it easy to integrate location and communication. This is difficult for normal radio. UWB radio has excellent anti-interference capabilities that can provide accurate positioning indoors and outdoors.

UWB accurate positioning principles are shown in Fig. 8.1.

Currently, there are many manufacturers who have developed and introduced UWB chips, application development platforms, and related devices represented by Ubisense, Timedomain, MSSI (later acquired by Zebra) companies.

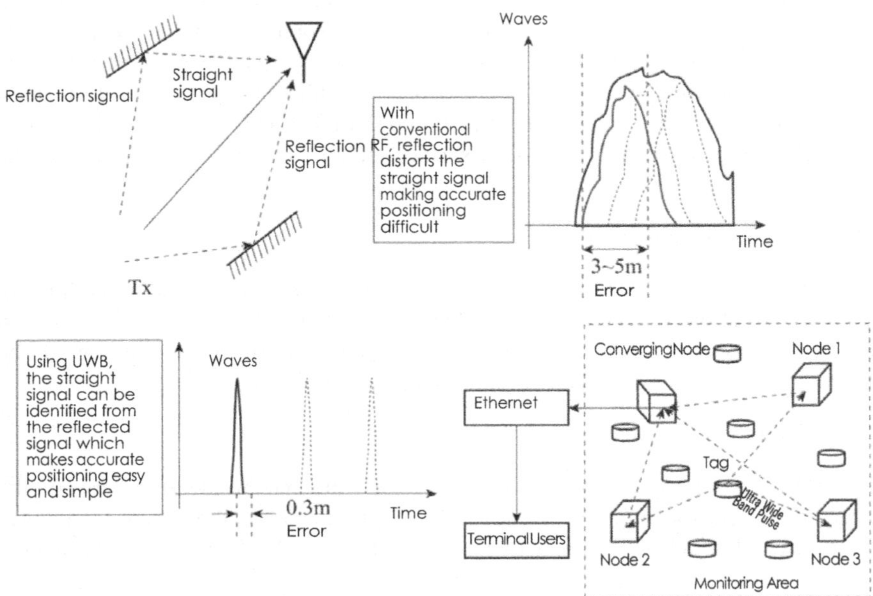


**Fig. 8.1** UWB accurate positioning principles diagram

## 8.2 Overall Architecture Design of the System

At present, radio positioning commercial projects based on UWB are few. Lack of successful experience can be used for reference. According to related materials, UWB accurate positioning systems generally have 4 components: Positioning Sensor, Active Location Tag, Timing Hub Distributors and Positioning Server (including positioning platform and data interface). In the system, the timing hub distributor is responsible for the clock synchronization of all sensors. The positioning tag actively transmits UWB pulse signals to the senors, and sensors adopt the TDOA/AOA positioning algorithm to conduct an analysis of the tag position after the signals are received and then transmit it to the positioning server by a wired Internet. The positioning engine in the server performs data calculation, filtering and optimization with output through a standard data interface. UWB positioning units can realize seamless cellular connections, thus expanding the positioning space infinitely. Positioning tags could move freely at any unit and be analyzed by positioning platform software. The target path positioned is displayed in a virtual dynamic three-dimensional effect.

The main components of the UWB accurate positioning system include the following:

(1) Positioning sensors

UWB positioning sensors include an antenna array and UWB signal receiver. The real location of positioning tags can be calculated by detecting UWB signals transmitted by positioning tags. In the operating process, each sensor measures the direction angle and AOA (angle of altitude) of UWB signals. While the TDOA must be measured by a pair of sensors, both sensors are deployed with time synchronization lines. This unique integrated measurement technology of TDOA/AOA can build flexible and powerful positioning systems. A surface antenna is commonly used in UWB systems.

(2) Active positioning tag

An active positioning tag is relatively small and can be placed on asset equipment and transportation. In addition, it can obtain 3D dynamical positioning information with an accuracy of 15 cm. Tag supports low-power sleep mode. The tag can be activated immediately by a built-in motion detection sensor.

(3) Timing hub distributors

The TOA positioning method requires accurate time synchronization between the target node and the reference node, and the TDOA positioning method requires accurate clock synchronization between the reference nodes. Timing hub distributors are responsible for all sensors' clock synchronization.

(4) Positioning server

Real-time positioning servers can obtain data from positioning sensors, active positioning tags and other RTLS sensing systems. Users can easily deploy the monitoring sites and coordinate the positioning network in real time and with high accuracy to obtain the positioning performance. This can dynamically virtualize the scene in real time.

## 8.3 Ideas for Designing System Functions

Since these ideas are similar to the local area positioning function in the previous chapter, this section will not be repeated.

## 8.4 Problems and Prospects

Whether prison radio positioning requires precise positioning at the centimeter level remains to be discussed. There is no doubt that precise positioning is under active technical development. In the case of UWB technology, positioning accuracy around a depth of 10~30 cm is enough to meet the requirements of personnel positioning. Of course, the cost of UWB deployment may be a key constraint.

Due to the early development of traditional radio frequency technology, various applications are relatively mature, while UWB positioning technology is a high-precision positioning technology that has just been developed in recent years. Its appearance has made up for the technical gap of high-precision positioning in traditional wireless technology.

As shown in Table 8.1, UWB positioning technology is far ahead of other systems in terms of positioning accuracy, refresh rate, and other technical parameters. Active RFID, ZigBee and other positioning technology and positioning effects are based on

**Table 8.1** Performance comparison of several radio positioning technologies

| Positioning technology | Accuracy (m) | Refresh rate | Anti-interference | Difficulty of deployment |
|---|---|---|---|---|
| UWB | 0.1~0.3 | 0.004~400 Hz Adjustable | Good | Complicated |
| CSS | > 2 | 0.1~2 Hz Fixed | OK | Complicated |
| ZigBee | 2~5 | 0.5~2 Hz Fixed | Vulnerable | Easy |
| Active RFID | 2~5 | 0.05~0.5 Hz Fixed | Vulnerable | Easy |
| WiFi | 2~5 | 0.05~0.5 Hz Fixed | Vulnerable | Easy |

simulated points in a two-dimensional environment. Meanwhile, UWB with three-dimensional positioning capabilities can achieve coordinate-level high-precision positioning, but considering the cost, it is still difficult to promote in prisons.

# Chapter 9
# Prison Cloud and Big Data Application—Auxiliary Decision Making Support System

## 9.1 System Overview

To achieve the government goals of ensuring the safety of supervision, improving the level of penalty correction, and promoting fair and honest law enforcement, a prison needs to conduct comprehensive monitoring and in-depth analysis of the entire prison's operation. The usual method is to analyze relatively static business data and the relatively dynamic Internet of Things supervision data over a certain period. From these results, we can judge whether every business aspect in the prison is running normally.

To fully explore the value of prison business data and transform the data into supporting information that can assist prison managers in making decisions, it is first necessary to fully integrate existing business systems and the perception system information resources of various prison departments and aggregate the information into the prison big data center. This provides comprehensive, accurate, and convenient information services for prison managers to make decisions. This can make full use of various professional systems and intelligent analysis models of various departments and extract the key performance indicators (KPI) based on data mining, statistical analysis, forecasting and early warning. Then, by evaluating the information and by visualizing and dividing the KPI into different topics for clear presentation, improved decision-making information service can occur, that is, to build an "auxiliary decision-making support system in the prison".

The topics most concerned by current prison law enforcement managers are shown in Table 9.1.

Decision Support System (DSS) is used to assist decision makers in identifying problems through data, models and knowledge; clarify decision goals; establish or modify decision models; provide various alternatives; evaluate and optimize various options; and create a human-machine system for semistructured or unstructured decision-making through human-computer interaction. The decision support system emphasizes the support of management decision-making rather than the automation of decision-making.



P. Sun, *Smart Prisons*,
https://doi.org/10.1007/978-981-16-9657-2_9

**Table 9.1** Prison management topics

| Topic name | Topic subitems | KPI |
|---|---|---|
| Supervision safety topic | Emergencies | Number of incidents, incident type, occurrence location, occurrence time, incident level, disposal plan |
| | Case in the prison | Number of cases, type of cases, level of charge of suspects, level of cases, rate of cases and detection rates, rate of escape of criminals, abnormal deaths of criminals |
| Criminal management topics | Penalty enforcement | Sentence commuting rate, parole rate, temporary operation outside prison (out-of-prison medical treatment) rate |
| | Education correction | Completion of corrective education, rate of skills certificate obtained, conversion rate of stubborn criminals, rate of recidivism |
| | Prison labor | The completion of prison labor and the change of labor income |
| | Medical health | Disease incidence rate, epidemic situation probability |
| | Logistics support | Purchase rate, adequacy rate and obsolescence rate of clothing, food and linens |
| Civilian police management topics | Team building | Age distribution of civilian police, distribution of civilian police ranks, distribution of civilian police academic qualifications, rate of change in civilian police positions |
| | Quality of law enforcement | Accused rate, reported rate, law enforcement transparency and prison affairs openness |
| | Law enforcement efficiency | Unqualified business ratio |
| Performance improvement topics | Equipment performance | Improve the key performance indicators of the system and equipment and increase the utilization rate |

(continued)

**Table 9.1** (continued)

| Topic name | Topic subitems | KPI |
|---|---|---|
| | Green | Comprehensive environmental performance, green energy saving and emission reduction, and centralized energy consumption management and control |
| | Police effectiveness | Integrate police resources, improve police efficiency, and improve service mechanisms |
| Special topic on crimes analysis | Criminal group structure | Gender ratio, age ratio, geographical division, length of sentence, type of crime, graded punishment |
| | Supervision of severe charges | Inquiries about the information of severely charged criminals, the statistics of the positions of the severely charged criminals, the category statistics of the severely charged criminals, the statistics of the dangerous level of the severely charged criminals, the statistics on the age of the severely charged criminals, the statistics of stubborn criminals, and the statistics of abnormal criminals |
| | The ratio of the police | Area-level police-criminal ratio, prison-level police- criminal ratio |
| | Psychological correction | Early warnings of psychological evaluation, psychological clinic, and psychological treatment |
| | Risk composite index | Comprehensive indicator of early warning of risk situation |
| | Public opinion about the prison | Early warning of public opinion about the prison |

The decision support system is generally composed of a human-machine interface/interactive language system, a problem processing system and a database system, a model library system, a method library system, and a knowledge base system with the following characteristics (Fig. 9.1):

- Integration. Decision support systems need to organically combine system analysis, operations research methods, computer technology, knowledge engineering, artificial intelligence, etc., under the premise of problem focus and use their respective advantages to realize the integration of decision support processes.

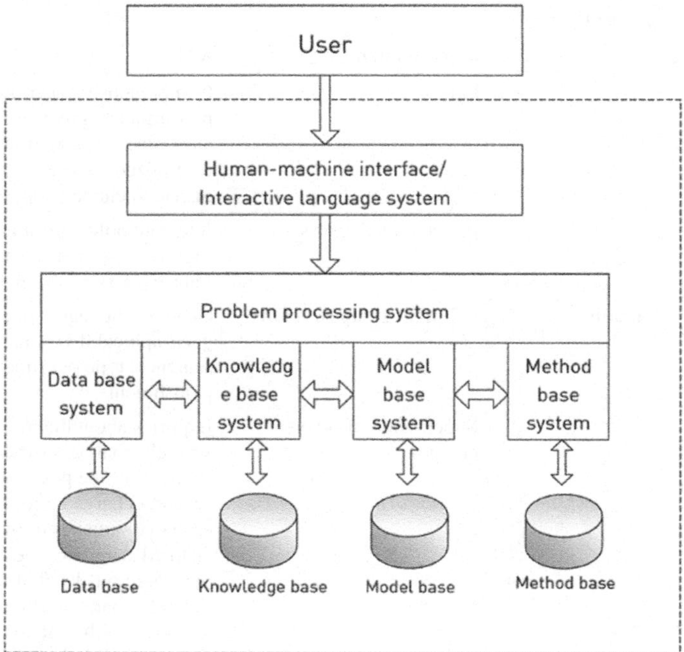


**Fig. 9.1** The structure of the decision support system

- Interactivity. The core of the decision support system is human–machine interaction. This is done to help decision-makers deal with semistructured and unstructured issues, clarify goals and environmental constraints, generate decision-making plans and conduct comprehensive evaluations of decision-making plans.
- Intelligence. Decision support systems need to use knowledge engineering and artificial intelligence methods and tools when dealing with problems that are difficult to quantitatively analyze.

Since the concept of decision support systems was put forward in the 1970s, decision support systems mainly consist of a model library, database and a human-machine interaction system, also known as the first stage of a decision support system. In the early 1980s, a knowledge base and methodology were added to the decision support system. In the late 1980s and early 1990s, with the development of neural networks and machine learning technologies, decision support systems began to be combined with expert systems (expert systems, ESs) to form an intelligent decision support system (IDSS). In the mid-1990s, new technologies such as Data Warehouse (DW), On-Line Analysis Processing (OLAP), and Data Mining (DM) appeared. Mining, model libraries, databases, and knowledge bases (a knowledge base based on deep learning) are now combined to form a more advanced form of comprehensive decision support system (synthetic decision support system, SDSS).

## 9.2 Overall System Architecture Design

The prison auxiliary decision-making support system relies on the prison big data center to perform statistical analysis and data mining on the basis of the criminal information database, the police staff information database, the prison management information database, the Internet of Things perception information database, and other data sources to extract and display the key information hidden in the data to form a statistical analysis report covering security supervision, criminal management, police management, efficiency improvement, crime situation analysis and other topics. This could be supplemented by other forms of display, such as radar charts, histograms, pie charts, graphs, and reports, to provide support for leaders to aid decision-making.

The prison auxiliary decision support system adopts a layered architecture design, and its overall architecture is shown in Fig. 9.2.

The overall logical architecture of the system is composed of a user layer, business application layer, application support layer, prison big data center, data source,

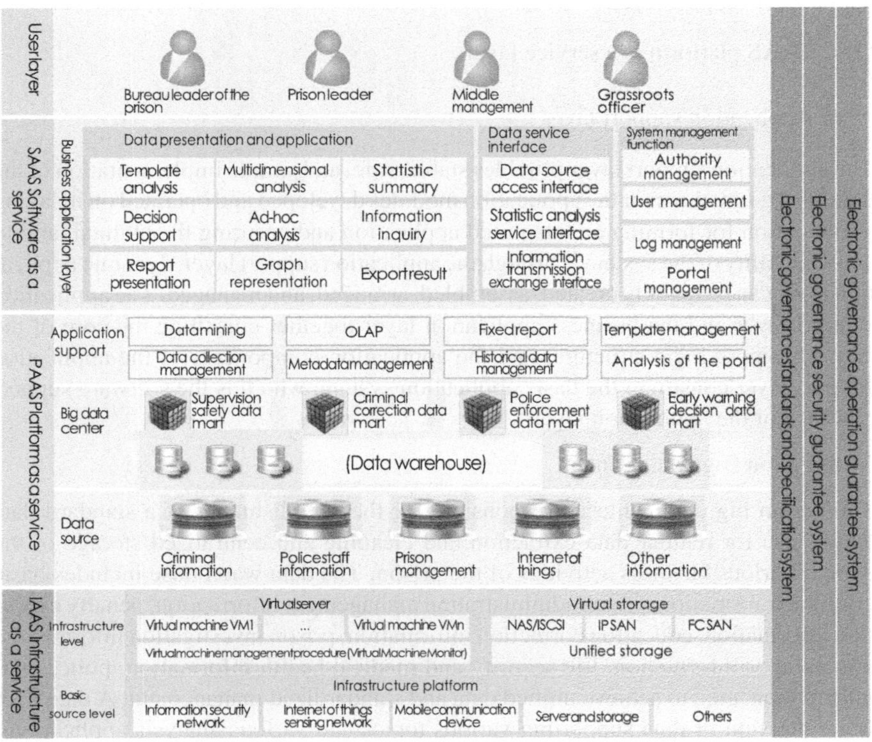


**Fig. 9.2** The overall architecture of the prison auxiliary decision support system

infrastructure layer and basic resource layer from top to bottom. The entire architecture focuses on the infrastructure service layer (including the basic resource layer and the infrastructure layer) and the data layer (including the prison big data center and data source) as the backing, with the application support layer and the business application layer as the core, providing comprehensive services for users to support decision-making services.

Summary description of each layer of business logic structure:

(1) User layer

Through unified user management for unified identity authentication, a single sign-on is realized. All kinds of users log in to the corresponding portal to access the information and business systems permitted by the user's authority.

(2) SAAS software is a service layer (business application layer)

The business application layer is the core of the entire business logic structure. This layer provides data display and application, data service interface, and system management functions by using the application component resources of the application support layer.

(3) PAAS platform is a service layer

◎ Application support layer

The application support layer provides stable, safe, and simple implementation of the underlying functions for the application modules developed and operated on it, laying a foundation for forming an integrated application and ensuring the maintainability and scalability of the system. Through the application support layer, dynamic application logic can be quickly created, assembled, arranged, and managed. The application support layer and the business application layer together constitute the core of the entire business logic structure, and the application components of the application support layer constitute the basic application system, which is the software support platform of the application layer.

◎ Prison Big Data Center

The prison big data center is responsible for the establishment of a standard data warehouse for regular data extraction and cleaning and centralized storage of the data of various business activities of the prison. The data warehouse includes basic criminal information, prison administration management information, penalty execution information, education correction information, prison investigation information, prison labor information, life security and medical health information, police staff information, etc., to achieve unified data and standardized management. A data mart is a collection of data supporting various topics and various analysis applications, including a data mart for supervision and safety, a data mart for criminal correction, a data mart for police enforcement, and a data mart for early warning and decision-making.

◎ Data source

The data source of the system is a database summarized from each prison to the data warehouse, including a criminal information database, police staff information database, prison management information database, perception information database of the Internet of Things, etc., providing comprehensive data source support for the prison big data center.

(4) IAAS infrastructure service layer

The infrastructure service layer mainly includes the infrastructure layer and the basic resource layer. The infrastructure layer includes prison private cloud infrastructure services based on virtualization-virtual servers, virtual storage, etc. The basic resource layer includes system software and hardware, security equipment, networks and other infrastructure platform resources required for system construction.

## 9.3 Design Ideas of the System Function

The functional design of the prison auxiliary decision-making support system is follows:

(1) Global decision analysis function

The overall decision-making analysis mainly serves prison macrodynamic analysis and aids leaders in decision-making. The prison auxiliary decision-making support system can perform statistical analysis and display the overall supervision business, using statistical icons to comprehensively analyze various business lines, such as prison penalty enforcement, education correction, prison labor, medical health, and logistics (clothing, food, quilt).

◎ Data mining and analysis

This uses mathematical statistics and artificial intelligence analysis methods to analyze data in the data warehouse, make inductive reasoning, dig out potential information and laws from data, and use intuitive statistical charts and other results to help leaders understand the prison macrodynamic situation, predict the risk of early warning, and provide accurate decision making.

◎ Data usage

Data usage provides functions such as information query, statistical summary, graphical display, data conversion and output, which provides support for the use of shared data.

The main functions of data usage tools include information query, statistical summary, graphical display, data conversion and output, data access interfaces, and information resource exchange interfaces, as shown in Table 9.2.

**Table 9.2** Main functions of data usage tools

| Module | Basic functional requirements |
|---|---|
| Information query | Supports multiple forms of information query methods such as single query, combined query, subject query, fuzzy query, full-text search, quick query, coding query, etc |
| Statistics summary | Supports numerical calculations and summary calculations commonly used in database applications, mainly including: fixed calculations, summary calculations, instant calculations, etc |
| Graphical display | Supports display of query and statistic results in a variety of ways such as histogram, pie chart, line chart, multidimensional histogram, etc., and supports the display rules of the proportion and color, etc. of the graphics definition |
| Data conversion and output | Supports output query statistic results according to user needs, support multiple file formats, support single table output and combined output |
| Data access interface | Supports the customization of special data access interfaces to meet the needs of using targeted information resources |
| Information resource exchange interface | Provides an interface with the information resource exchange system to facilitate point-to-point exchange of certain data through the information resource exchange system |

(2) Topic decision analysis function

Topic decision analysis mainly serves the topic analysis of business analysts. Similar to the global decision-making analysis function, it is based on the statistical analysis of data to form multiple types of thematic decision-making analysis functions, such as supervision safety, criminal correction, police enforcement, and early warning decision-making.

For example, this system can display data on the supervision of prisoners under serious charges in prisons, as well as data on the efficiency of law enforcement by the police. The data on criminals and police can be further analyzed, including net age structure analysis and police-prison ratio analysis.

Similar statistical methods can be applied to all aspects of prison management to cover various businesses, positions and types of personnel, forming a comprehensive, accurate and reliable decision-making support method.

(3) Information resource management module

◎ Screening and processing of old data

In accordance with the requirements of prison information regulations, specialized tools are used to extract and clean old data, qualified data, and unqualified data; these are classified one by one and transmitted regularly and stored uniformly.

◎ Historical data management

This regularly gathers, processes and analyzes data, builds indexes, searches on demand, automates statistics, etc.

◎ Data reuse

According to the needs of other business systems, a variety of different data acquisition modes are provided to improve data utilization.

◎ Data timeliness

To ensure the validity of the data, the frequency of data extraction and cleaning can be adjusted as needed.

## 9.4 Problems and Prospects

A decision support system can use shared decision resources (data, models, knowledge) to assist in solving various decision-making problems. The data technology revolution in the era of big data has brought new development opportunities for decision support systems. These systems can support decision-makers in conducting global decision-making through the online collection and real-time analysis of comprehensive, structured, and unstructured real-time data and historical data, especially behavioral characteristic data hidden behind appearance data.

In specific applications to prisons, due to the interrelationship of various factors, the complex decision-making environments and inaccurate and incomplete data has caused great difficulties in the construction of decision support systems.

The main application limitation of the prison decision support system lies in the current prison data scale and application level (except for video applications, the construction of prison sensor networks and various business systems that have not yet been completed). It is urgent to start from data, deepen the insight and mining of prison data, and promote the realization of the value of data governance. This can be done by focusing: on the work content of the data governance domain; governance of data acquired from the various business systems and perception systems of the prison; on the construction of data standards, data access, data processing, data models, data quality, data security, data services and value realization; and forming all kinds of original libraries, topic libraries, knowledge bases and business libraries to achieve the purpose of data fusion and knowledge accumulation.

# Chapter 10
# Prison Cloud and Big Data Application—Criminal Situation and Recidivism Risk Early Warning System

## 10.1 System Overview

At present, the monitoring and evaluation of the risk of prison criminals is mainly through artificial means, using the subjective questionnaire method of social psychology. The evaluation scene is single, and evaluation accuracy is easily influenced by the subjective intention or unintentional cover-up of the criminals. In a prison's confrontation environment, criminals' behavioral clues are highly concealed, and their preventions are strong. Potential psychological activities and behavioral intentions are difficult to explore through scales and traditional questionnaires. The evolution of prison crimes and the accuracy of recidivism risk assessment depend entirely on the police's level of specialization, which causes uneven quality in artificial evaluation. Some existing risk assessment systems require the police to input a large amount of crime data based on artificial judgments to realize early warning and prediction (the results obtained by the system's "intelligent analysis" are still based on artificial judgments), which not only fails to reflect the benefits of intelligent systems that improve quality and reduce burdens but also requires a large amount of artificial data entry work that increases the burden on the police (Fig. 10.1).

The evolution of crimes and the risk of recidivism (recidivism rate) of criminals during prison sentences are affected by a variety of factors, such as social, psychological, and specific situations. Traditional artificial psychology/behavior risk assessment and correction strategies have difficulty meeting the demands of being dynamic, real-time and accurate. Therefore, it is necessary to establish a multidimensional reform quality evaluation system based on the "resocialization" model of criminals serving sentences throughout the entire cycle and construct a situation analysis model for the evolution of crimes that covers the elements of the crime, risk elements, risk signs, root causes, etc. On the basis of psychology/behavior and a correction strategy matching deduction, a quantitative analysis system for the evolution of crimes should be established for different stages from entering prison to leaving



P. Sun, *Smart Prisons*,
https://doi.org/10.1007/978-981-16-9657-2_10

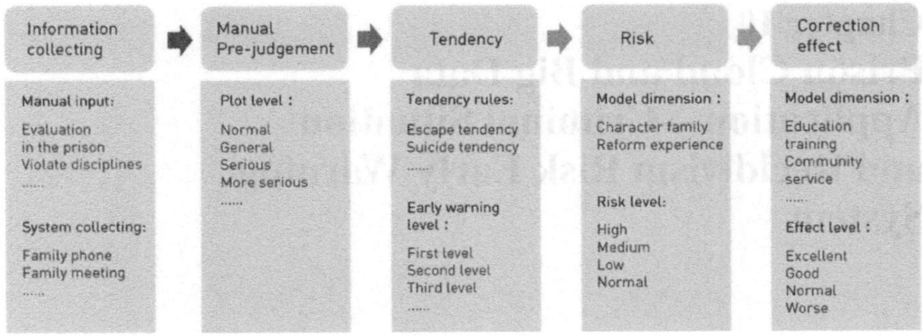


**Fig. 10.1** Criminal risk monitoring and assessment procedure

prison, including constructing a multidimensional associated network that drives relationships of human beings, objects and events in the social/personal relationship space and predicting the risk of recidivism in each stage.

## 10.2 System Overall Architecture Design

Due to the strong dynamics of criminals' psychological and behavioral changes, it is difficult to make timely adjustments and follow-up of correction strategies. To solve problems such as a fuzzy understanding of the evolution of psychology, behavior and correction strategies, the evolutionary situation of crimes, and the low accuracy rate of predicting the risk of recidivism, a crime quantitative indicator system and a quantitative analysis of the crime evolution situation system need to be built. It should be based on the full-period multidimensional criminals' data of "imprison--- serve sentence---discharge" (criminal basic data, penalty correction data, physiological- biochemical-psychological-behavioral data, correction strategy data, the internet of things perception data, business collaboration data, etc.) The interpersonal relationship and the community relationship in prison can be integrated. With these features and associated relationships, people can build a model about the evolution of crimes, which analyzes the quantitative correction strategy of criminal psychology and behavior. For recidivism risk prediction and early warning, multielement, multidimensional, multilevel risk data calculations and evaluations are established to provide support for real-time monitoring of prison infringement and dynamic adjustment of correction strategies.

The situation and recidivism risk early warning system are shown in Fig. 10.2.

(1) To construct an analysis model for the evolution of crime

Relying on multilayer complex network analysis, evolution and modeling theories, people can combine the knowledge of criminal evolution to build a prison where a

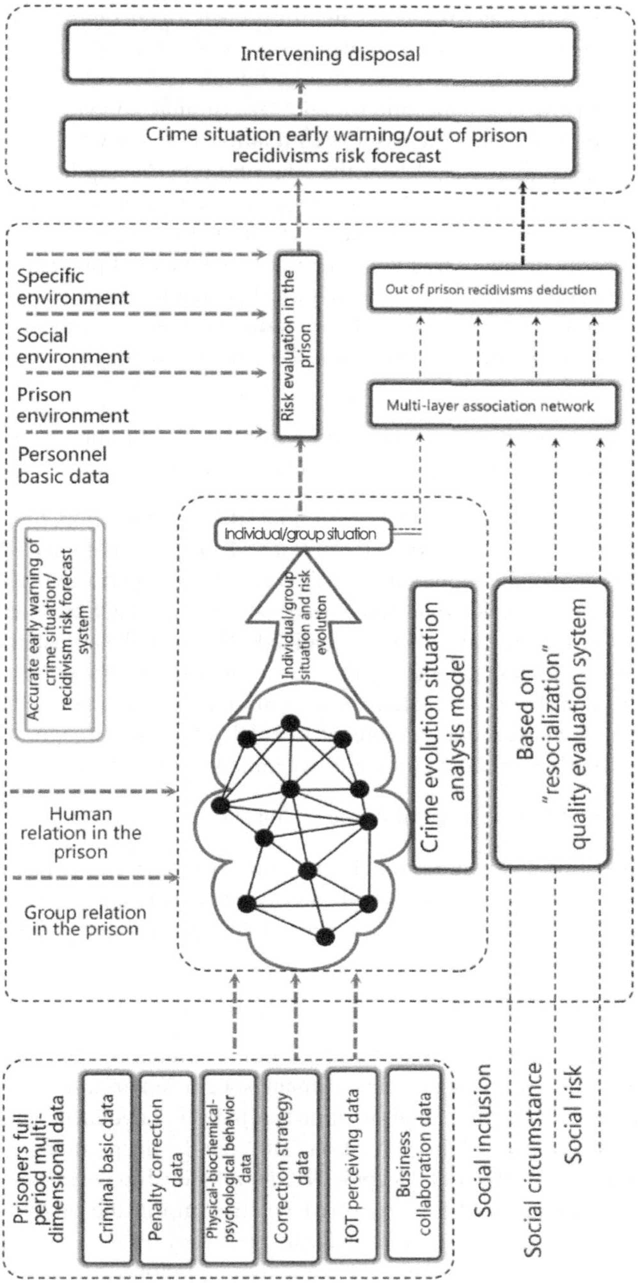


**Fig. 10.2** Situation of crimes and the overall architecture of the recidivism risk early warning system

"location---event---person---behavior" multilayer heterogeneous network with "people" is the core. This creates a collection of people's attributes and their relationships, forms a multilevel, specific knowledge representation framework and demonstration method, and constructs a conceptual model of criminal knowledge in the prison complex system. It analyzes the associated relationships between different objects at the same level of the network and the coupling relationship between objects at different levels of the network. It also integrates time factors to model the time-varying relationship between prison places, events, people, behaviors, and risks. The model analyzes the interaction of the influence mechanism of prisons' security risk, personnel behavior patterns and psychological dynamic principles with different dimensional characteristics. It allows for quantitative analysis of the evolution trend of individual/group situations of criminals. The analysis model of crime evolution is shown in Fig. 10.2.

(2) To establish a quantitative indicator system for criminal situations and a quantitative analysis system for criminal situation evolution

On the basis of this analytical model of criminal situation evolution, people can build a chain reasoning knowledge model for prisons' decision-making based on key risk incidents. This covers the individual's basic data, physiology/location data, psychological/behavioral deduction data and correction strategy matching data. These data are extracted by features and then normalized. The model is clustered according to the frequency domain characteristics of criminals' historical risk characteristics. Through neural network calculations, multielement, multidimensional correlation panoramic analysis and comprehensive risk situation perception, projections of criminal individual/group evolution are obtained. As shown in Fig. 10.3.

◎ Extraction of key features of crime evolution

Data, such as criminal psychology, behavior, and correction strategies, are all multi-source and multidimensional. The weight and cycle of data influence on the evolution of the situation are different, so it is necessary to extract linearly independent key characteristics of these data.

◎ Quantitative analysis method of the crime evolution situation

According to the key characteristics of the criminal situation, basic personnel data, prison environment, overall social environment, specific environment and other data, a A quantitative criminal individual situation risk index is calculated using this analytical model of the crime evolution situation. Through the deep graph neural network (Deep Graph Neural Networks, DeepGCN), the relationship between individuals is automatically obtained, and the group risk relationship in the monitoring interval is modeled; the group situation risk index is predicted by the individual situation risk index. Since the input data is time sequential, the model can have the dynamic characteristics of individual group risk changes over time.

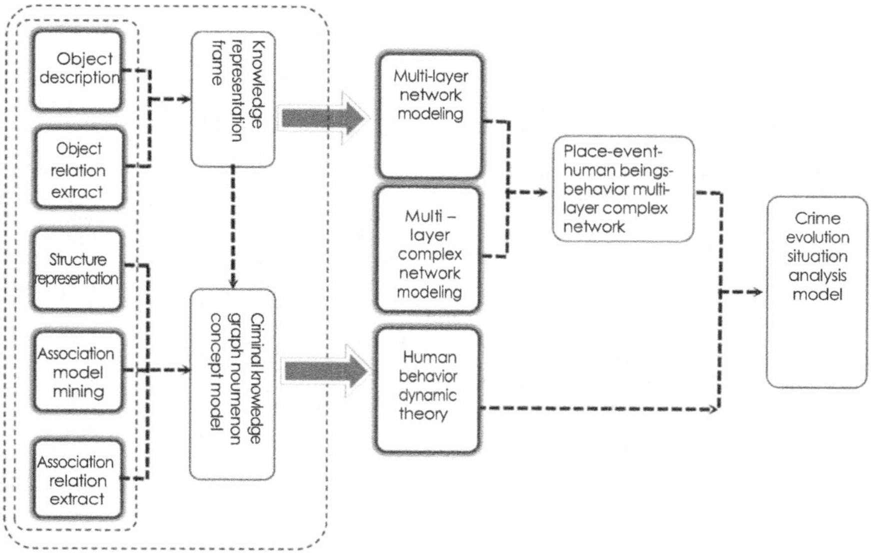


**Fig. 10.3** Constructing an analysis model of crime evolution

(3) The risk prediction and early warning system for recidivism is based on the "resocialization" quality evaluation system

According to the data on the individual evolution of criminals, using time and space dimensions as the baseline, people can construct a multilayer association network based on the psychological behavior, social integration, social situation, and social risk elements of released prisoners. This network is based on the expressive mechanism of recidivism risk and the driving mechanism between recidivism risks, based on the traditional socialization model of people, combined with the quality assessment of the "resocialization" of the criminals' full-period sentence (i.e., reintegration into society to become a law-abiding citizen). It measures the recidivism of criminals after being released from the prison. Through this multidimensional process, one can project the early warning of the risk of recidivism in the short, medium and long term, as shown in (Figs. 10.4 and 10.5).

◎ Short-, medium-, and long-term recidivism risk prediction and early warning method

The traditional practice of projecting the risk of recidivism after the release of prisoners is based on the actual performance of the criminal in prison, with "mental health assessment" and "recriminal possibility assessment" carried out before release. Recidivism tendencies and the types of help and education are studied and comprehensively researched. However, there are subjective differences and contingencies, and the actual recidivism effect is difficult to evaluate, especially when the social

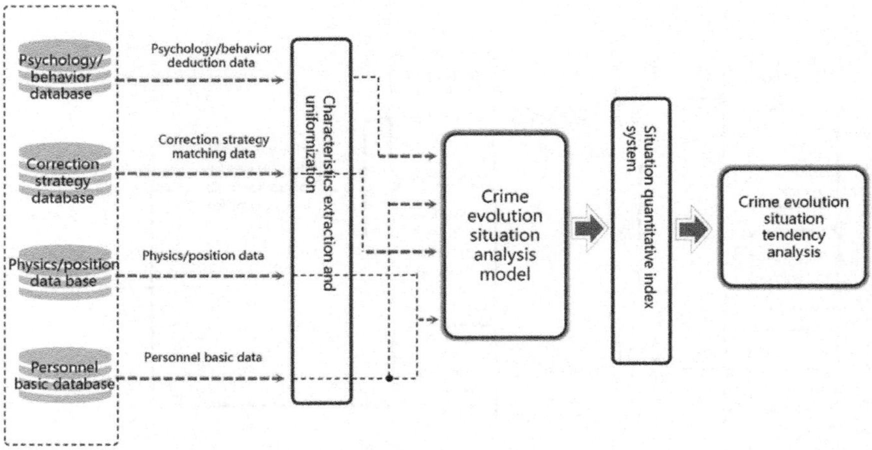


**Fig. 10.4** The structure of the quantitative analysis system for the evolution of the crime situation

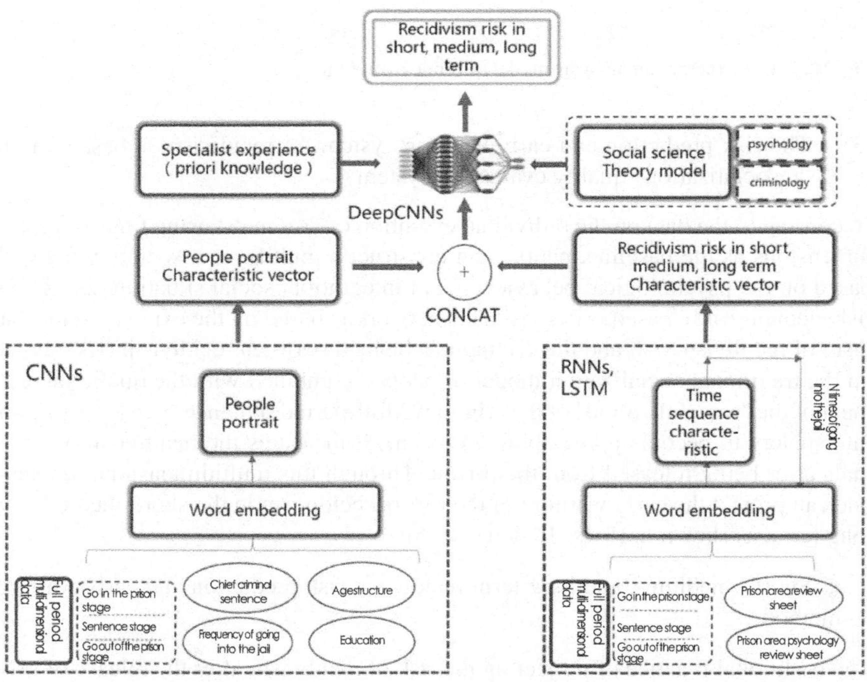


**Fig. 10.5** Structure of the recidivism risk prediction and early warning system

environment and support for released prisoners are often unpredictable. Additionally, data collection is difficult, so it is very difficult to predict the risk of recidivism in the short, medium, and long term.

By using machine learning algorithms to learn training scale scoring rules and based on the multidimensional data of the criminals' full period in prison and for criminals who need prison evaluations, convolutional neural networks (CNNs) can help to extract crime types and age structures, education levels and chief criminal sentences. These are obviously related to characteristic recidivism data and can help to form characteristic vectors of criminal character. Modeling the characteristics of the criminals, using data on times of criminal recidivism, integrating the characteristic information of the time series, and using recurrent neural networks (RNNs) and long short-term memory (LSTM) can help to extract the risk of recidivism for criminals across different times of imprisonment to generate short-, medium-, and long-term recidivism risk features for each released prisoner. The integration of character portrait features and short-, medium-, and long-term recidivism risk features as well as expert experience and the knowledge of social science theoretical models (psychology, criminology) can also help. The final step is to apply deep graph neural networks (deep graph neural networks, DeepGCNs)---to model spatial characteristics when social environmental adaptability and social support system data cannot be determined. This improves the accuracy of the quantitative prediction of criminals' short-, medium- and long-term recidivism risks and can enhance the efficiency and intelligence level of recidivism risk assessment.

## 10.3 Design Ideas of System Function

The functional design of the recidivism risk early warning system is as follows:

(1) Analysis of overall prison data

These data can contribute to overall analysis at the prison level, such as the changing trend of the number of criminals, the distribution of criminals' household registration, the number of criminals in each prison, and the overall structure of criminal characteristics (using cluster analysis of several dimensions, such as category, sentence, remaining sentence, age, education level, occupation before arrest, marriage status, type of detention, number of previous convictions, hierarchical treatment, etc.), and analysis of high-risk criminals, etc., using the overall data of criminals in the prison.

(2) The perception of overall crimes

This reflects on the security of the entire prison, the state of criminality, the ratio of real-time police and prisoners, the serviceability rate of the Internet of Things, security guard facilities, and multiple factors in five main dimensions: place, event, person, behavior, and risk. These data help perform predictive perception of crimes in the prison and supervision areas.

(3) Portraits of people

◎ Individual portrait of the police

This displays basic information on the police, including duty, law enforcement, attendance, shift scheduling, and other dimensions.

◎ Individual portraits of criminals

This displays basic information on individual criminals, punishment correction information, scoring assessment information, meeting and family telephone information, information on consumption in prison, information on commutation and parole, etc.

(4) Graph of the relationships of criminals

The criminal relationship graph includes relationships between the police (in charge of the police) and prisoners, the relationship between prisoners (in the same prison area and room, fighting, cocriminals, etc.), and meeting relationships (relatives, nonrelatives, lawyers, public security, procuratorate, courts, etc.), calling relationships (relatives, nonrelatives, etc.), income and expenditure relationship and more.

(5) Quantitative assessment of individual/group comprehensive risks of criminals

◎ Criminal personal information includes the criminal's personal information, personality characteristics and fluctuations in psychological behavior.
◎ Criminal danger level and quantification value of comprehensive risk. Charts are used to distinguish different levels of risk and use different colors to present low, medium, high, and extremely high risk levels.
◎ Criminal danger category. By combining the hazard level chart to distinguish different types of hazards, this chart presents the three-dimensional hazard categories of suicide, escape, and assault. The location of individual data identifies different hazards categories.
◎ Adjust the classification, separation, and classification of treatment. Criminals of different levels of danger are detained at different levels of prison security and adjusted according to control measures.
◎ Disposal of abnormal behavior. Prison administration, prison investigation and other departments provide abnormal behavior case reports to form a case database.

(6) Risk assessment of recidivism

◎ (Before prison release) Basic criminal information. This includes the criminal's personal information and the full-period risk quantification information of "entry into prison--serve sentence--leave prison".
◎ (Before prison release) The criminal's risk level of recidivism. This is divided into low, medium, high, and extremely high risk tendency levels. An automatic warning prompts extremely high-level risk.

◎ Categories of help and education target. According to the classification of recidivism risk tendency, criminals with low and medium risk are listed as general help and education targets, and criminals with high and extreme high risk are listed as key help and education targets.
◎ Disposal recommendations. According to the possibility of the criminals committing a crime again after being released from prison, penal enforcement department proposes whether to apply parole, temporary execution outside of prison, etc. The education and correction department proposes the arrangement and education of the criminals after release from prison.

## 10.4 Problems and Prospects

To resolve the difficulty of rapid extraction and quantification of risk characteristics that exist in traditional risk assessment tools, the large subjective differences and contingency, and the lack of professional police for risk assessment, more efficient models are advancing with the informationization of prisons. Static and dynamic data collection in real time has become increasingly abundant, and this has created conditions for better implementation of automatic and intelligent criminal risk assessments.

In the current information-based supervision environment, the design of risk assessment systems usually revolves around various events that occur during the full cycle of criminals that are used as scoring elements. The risk assessment system continuously collects real-time data that are used to construct weighted integral statistics to project danger levels and development trends of criminals. However, simple weighted integral statistics often do not take into account the correlation between various factors (for example, when criminals cry after receiving a family call that the risk assessment system has difficulty attributing to repentance or increased risk of suicide) and do not consider the prison itself as a constantly evolving ecological environment. So there is great possibility for misjudgment. In practical applications, the probability of identification error and false or missing reports in weighted integral statistics is relatively high.

# References


Alpaydin, E. (2009). *Introduction to machine learning*. Beijing: China Machine Press.

DoD Architecture Framework Working Group. (2010). *DoDAF2.02*. USA: the DoD of USA.

Duermeyer, K. (2004). *Methodology: From component business model to service oriented architecture*. IBM Press.

Erl, T. (2010). *Service-oriented architecture: Concepts, technology, and design*. Beijing: Science Press.

Federal Enterprise Architecture Program Management Office. (2005). *FEA consolidated reference model document*. http://www.whitehouse.gov/omb/egov/documents/CRM.PDF.

Feng, L. (2012). *Internet evolution*. Beijing: Tsinghua University Press.

Han, J., Kamber, M., & Pei, J. (2012). *Date mining concepts and techniques* (3rd ed.). Beijing: China Machine Press.

Hickman, L. J., Davis, L. M., Wells, E., & Eisman, M. (2010). *Tracking Inmates and Locating Staff with Active Radio-Frequency Identification (RFID): Early lessons learned in one U.S. Correctional Facility*. Santa Monica, CA: RAND Corporation. https://www.rand.org/pubs/technical_reports/TR786.html.

Hongyu, Y., & Suning, T. (2013). *Cloud computing: Systems engineering in the era of big data*. Beijing: Publishing House of Electronics Industry.

International Organization for Standardization ISO/IEC27001. (2005). 2005, Information technology-Security techniques-Information security management systems-Requirements.

Limin, S. (2005). *Wireless sensor networks*. Beijing: Tsinghua University Press.

Mayer-Schönberger, V. (2012). *big data: a revolution that will transform how we live, work, and think*. Hangzhou: Zhejiang People's Publishing House.

Negnevitsky, M. (2012). *Artificial intelligence: A guide to intelligent systems* (3rd ed.). Beijing: China Machine Press.

Sun, P., Zhang, H. (2012). *Prison internet of things*. Beijing & Wuhan: Tsinghua University Press & Huazhong University of Science & Technology Press.

Sun, P. (2013). *Community correction informatization*. Beijing & Wuhan: Tsinghua University Press & Huazhong University of Science & Technology Press.

Sun, P. (2014). *Smart prison*. Beijing & Wuhan: Tsinghua University Press & Huazhong University of Science & Technology Press.

TOGAF9.1. *White paper: An introduction to TOGAF Version 9.1*. http://www.opengroup.org/togaf/.

Turban, E. (2012). *Business intelligence: A managerial approach* (2nd ed.). Beijing: China Machine Press.




P. Sun, *Smart Prisons*,
https://doi.org/10.1007/978-981-16-9657-2

La Vigne, N., Halberstadt, R., & Parthasarathy, B. (2009). *Evaluating the use of radio frequency identification device technology to prevent and investigate sexual assault and related acts of violence in a women's prison.* Washington, USA: URBAN Institute. https://www.urban.org/sites/default/files/publication/30706/411972-Evaluating-the-Use-of-Radio-Frequency-Identification- Device-Technology-to-Prevent-and-Investigate-Sexual-Assault-and-Related-Acts-of-Violence-in-a-Women-s-Prison.PDF.

Virtualization and Cloud Computing group. (2009). *Virtualization and Cloud Computing*. Beijing: Publishing House of Electronics Industry.

Wei, R. (2012). *Internet of things security*. Beijing: Tsinghua University Press.

Yunhao, L. (2010). *Introduction to internet of things*. Beijing: Science Press.

Zachman, J. A. (1987). A framework for information systems architecture. *IBM Systems Journal, 26*(3).

Zhiyuan, L., Han, X. U., & Maosong, S. (2020). *Knowledge graph and deep learning*. Beijing: Tsinghua University Press.

Zhuo, Z. (2019). *Risk assessment and risk control for offenders* Beijing: China People's Public Security University Press.

The manufacturer's authorised representative in the EU is Springer Nature Customer Service Centre GmbH, Europaplatz 3, 69115 Heidelberg, Germany. If you have any concerns regarding our products, please contact ProductSafety@springernature.com

Printed and bound by CPI Group (UK) Ltd, Croydon, CR0 4YY
30/09/2026
02225264-0014